D1165900

HORSE POWER

HORSE POWER

A History of the Horse and the Donkey in Human Societies

JULIET CLUTTON-BROCK

Harvard University Press
Cambridge Massachusetts
1992

10 9 8 7 6 5 4 3 2 1

Designed by Gillian Greenwood

Library of Congress Cataloging-in-Publication Data
Clutton-Brock, Juliet.
Horse power: a history of the horse and the donkey in human societies / Juliet Clutton-Brock.
p. cm.
Includes bibliographical references.
ISBN 0-674-40646-X
1. Horses—History. 2. Donkeys—History. 3. Mules—History. 4. Equus—History. 5. Animals and civilization. I. Title.
SF283.C63 1992
636.1—dc20 91-20447
CIP

Phototypeset by Keyspools Limited, Golborne, Lancs, Great Britain
Printed and bound by Artes Graficas, Toledo S.A., Spain

D.L. TO: 185-1992

End Papers: Painting on elk hide
(National Anthropological Archive,
Smithsonian Institution, Washington)

Contents

Acknowledgements

My greatest thanks are due to Mary Littauer for her detailed comments, corrections, and help with the illustrations. I am most grateful to Richard Meadow for his helpful advice, to Isobel Smith for her masterly editing of the manuscript and to Joy Law for picture research and the index. My thanks are also due to Iain Bishop, Peter Jewell, and three anonymous readers for their comments and corrections. The figures and maps were drawn by Rebecca Jewell.

Preface

This book is about the very close association that human beings have had with the horse and the ass for many thousands of years. It begins by examining the natural history and behavioural patterns of these equids in the wild and describes their widespread distribution at the end of the last ice age, their gradual decline, and their near extinction at the present day. At the same time that the wild species were declining their domesticated counterparts were in an ascendance that led to their numbers and distribution far outstripping that of the wild species.

The expanding roles of the horse and the ass in human societies are explored through history until the steam engine was at its zenith, at the end of the nineteenth century, and mechanical power began to overtake horsepower. It may be argued that without the horse human history would have been entirely different. The great battles of the ancient world would have been mere civil squabbles, Alexander the Great could not have made his Asian conquests, and the Normans would not have invaded Britain in 1066. There would have been no Crusades and no foreign empires for, without fast transport and fast movement of goods, weapons, and food, invaders are powerless. In the Americas, the Incas and the Aztecs had managed to acquire and hold large empires without benefit of either horse or wheel, but they had no defence against the Spanish invaders who first arrived with their horses in AD 1492.

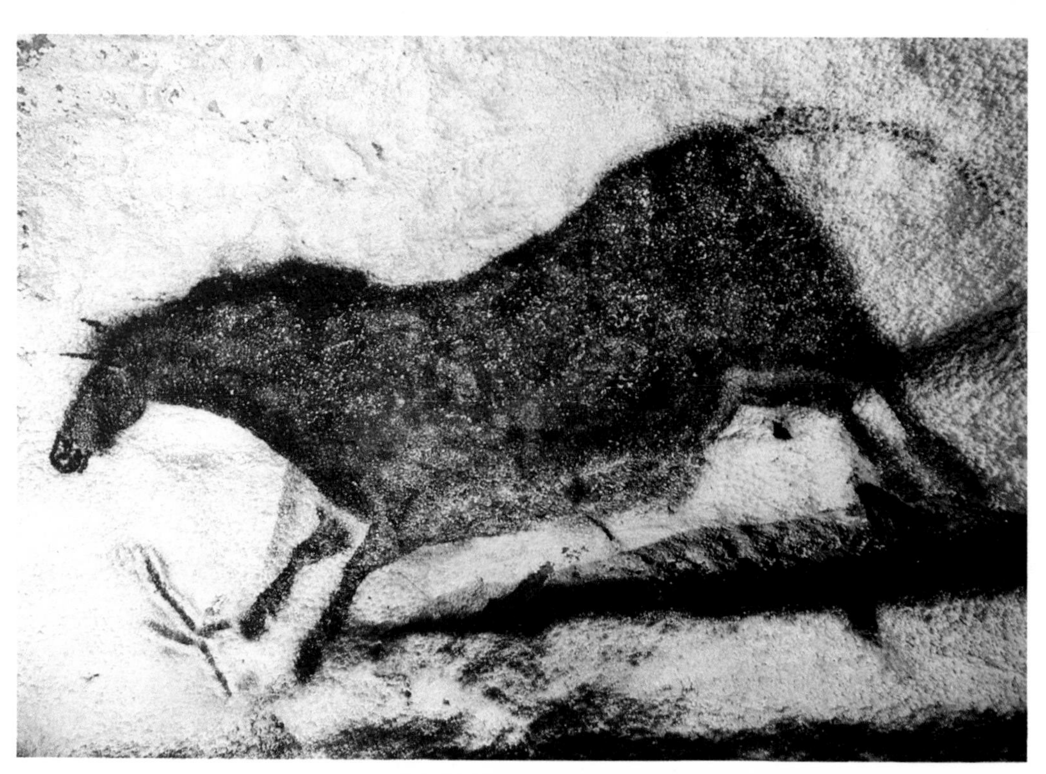

Introduction

The role of horsepower in the evolution of ancient civilizations

The earliest evidence for the horse being in some way a part of human culture comes from the cave paintings of the Magdalenian phase of the Upper Palaeolithic in France and Spain (Fig. 0.1). For the greater part of the long period since these works of art were painted, about 15 000 years ago, wild equids were hunted, right across the northern hemisphere, to provide meat and hides for human populations that were steadily increasing in numbers. Over-hunting, combined with climatic change, took its toll and the herds of wild horses began to dwindle. Equids were extinct in North America by 10 000 years ago (Mead & Meltzer 1984) and in Europe they were gradually pushed eastwards into central Asia where the last few Przewalski horses survived into this century.

By 9000 years ago, the increasing aridity of the climate in western Asia together with increasing pressure on the available supplies of wild plants and animals by expanding human populations most probably provided the impetus for new ways of obtaining food. People began to cultivate cereals and to domesticate goats, sheep and, rather later, cattle and pigs. At this period, as can be deduced from the absence of their remains in the fossil record, there were no wild horses (*Equus ferus*) in the region of western Asia where the domestication of livestock first took place. There were wild asses (*Equus africanus* and *Equus hemionus*) but perhaps because they were more difficult to restrain and to handle as herd animals than goats, sheep or cattle, asses were not included in the first wave of domestication, about 9000 years ago, and equids never became primary sources of meat.

It is not until around 6000 years ago that the remains of the earliest domestic horses are found on archaeological sites in the Ukraine, with a few scattered records from early sites as far west as central Germany (Anthony 1991; Glass 1989). At the same period, the earliest domestic donkeys begin to appear on wall paintings and in burials from Egypt and western Asia (Fig. 0.2). Perhaps, throughout prehistory, these equids were occasionally tamed and ridden, but there is no archaeological or osteological evidence for Kipling's explanation about their domestication by hunting people before the emergence of agriculture, as described in his famous *Just so stories*:

Fig. 0.1 The extinct European wild horse, one of many painted on the walls of the Lascaux caves, France, about 14 000 years ago. (Photo © Ronald Sheridan Ancient Art and Architecture.)

Fig. 0.2 Semite with his donkey. Tomb 3, Beni Hasan, *c.* 1900 BC (Photo reproduced by courtesy of the Trustees of the British Museum.)

When the Man and the Dog came back from hunting the Man said, 'What is Wild Horse doing here?' And the Woman said, 'His name is not Wild Horse any more, but the First Servant, because he will carry us from place to place for always and always and always. Ride on his back when you go hunting.'
Kipling 1902

Until very recently, archaeologists held the view that there was no evidence for the riding of the earliest domestic horses or asses. It was believed to be more probable that equids were laden with goods or harnessed to sledges and, later, to wheeled carts and chariots, with riding being a very rare event, as discussed in Chapters 4 and 6. However, horses cannot be moved about in any numbers without a mounted herdsman, so there is an inherent probability that they must have been ridden from the beginnings of domestication (Anthony 1991).

Human burials associated with the remains of asses and pieces of harness have been excavated from Sumerian sites in ancient Mesopotamia ('a country between rivers', now called Iraq) dated to around 4500 years ago. A model of a wheel found in Bikova in Bulgaria may date from as long ago as 5500 years, but there is no proof that the draught animal at this early period was the horse, for it seems certain that oxen preceded equids for traction.

Evidence for the early domestication of horses and asses comes from a large number of sources as, apart from the excavation of equid bones from settlement sites, an extraordinarily large number of burials, over several millennia and in many countries, have been discovered which contain complete equid skeletons, harness, and wheeled vehicles. In addition there are often models of horses and vehicles in graves which range in grandeur from those in the recently discovered tomb of the first Emperor of China (259–210 BC) at Xian to a small pottery model of a waggon from Syria (see Figs. 7.11 and 0.3). Then there are pictographs and cuneiform texts from Mesopotamia, Assyrian stone reliefs, wall paintings and hieroglyphs from ancient Egypt, as well as innumerable Celtic symbols. But perhaps most impressive of all is the treasure from the frozen tombs of the Scythians at Pazyryk, and from more recent excavations in southern Siberia (seventh to fourth centuries BC) of gold jewellery (often figuring horses), embroidered felt saddles, and elaborate harness.

By the first millennium BC the world was opened up to the horse-rider who could travel for the first time at a speed that far surpassed that of the ox-cart, or even of humans when running their fastest. In the fourth century BC Alexander the Great (356–323 BC) on his horse Bucephalus (Fig. 8.1) conquered two million square miles of the ancient world. He learned his horsemanship from the Scythians and from the writings of Xenophon (428/7–354 BC). From this time

Fig. 0.3 Pottery model of a waggon from a tomb at Hammam, Syria. 2200–2100 BC. (Photo Ashmolean Museum, Oxford.)

onwards the horse became of increasing importance in the art of war. Surprisingly, the stirrup was unknown until the first centuries AD and neither the Greek nor the Roman cavalry rode with stirrups.

Once the use of stirrups became widespread, mounted soldiers could wield the lance and bowmen could shoot from the saddle, with very little training. This meant that protective armour for the soldier and his horse increased in complexity and weight which, in turn, required the breeding of much larger horses. By Medieval times the heavy horse was beginning to appear and was taking over from cattle for ploughing and traction in northern Europe. This period can be justly called 'the age of the horse'. Radical changes came about not only because the use of the stirrup spread but also because of the invention of the horse collar, which meant that horses could pull loads with greater efficiency, and the introduction of the nailed horseshoe which came into universal use in the eleventh century.

The arts of warfare and hunting, chivalry, and indeed the whole structure of feudal society in the Middle Ages was built on the horse, and the distances covered by travellers were enormous. Armies of mounted knights from northern Europe were able to travel through the whole of Europe to the Levant, where Jerusalem was captured from the Muslims during the first Crusade in 1099.

In Asia, during the same period, Genghis Khan (1162–1227), with a cavalry of more than 30 000 men, each with two horses, broke through the Great Wall of China and then conquered all the lands from Korea to the Caspian sea, creating a vast empire which survived for three generations.

From the end of the Middle Ages, until mechanical power began to replace horsepower in eighteenth century Europe, the horse continued to be the essential pivot of civilization, providing the means of long-distance transport, agriculture, industry, and warfare. The politics of nations were swayed by battles won by the great soldiers, from Cromwell to Napoleon, who all depended for success on the strength of their cavalry. The point is neatly made by the seventeenth century nursery jingle:

> For want of a nail the shoe was lost,
> For want of a shoe the horse was lost,
> For want of a horse the rider was lost,
> For want of a rider the battle was lost,
> For want of a battle the kingdom was lost,
> And all for the want of a horseshoe nail.
>
> *Opie & Opie* 1951

The domestic donkey is slower than the horse, and, being descended from the wild ass of the hot deserts of Africa and Arabia, is also somewhat less viable in temperate regions of the world than the horse. Throughout history the donkey has fulfilled the role of pack animal for the trader and farmer in the Mediterranean regions, but perhaps most importantly the donkey has been sire to the mule.

Mules became an essential means of transport in the ancient world and remained so until the building of the railways. The Sumerians, 4000 years ago, may have been the first to breed hybrids between the domestic donkey (descended from the African wild ass, *Equus africanus*) and the wild Asiatic ass, *Equus hemionus*. Later it was found that even stronger hybrids could be obtained by breeding a male donkey with a female horse and this became the accustomed method of producing the most powerful and resilient baggage animals for peace and war.

In his book on the horse, published in 1891, Sir William Flower wrote:

> It is only in very recent times that the progress of mechanical invention has begun to supersede some of the uses for which the strength and the speed of the horse for many thousands of years have alone been available. How far this commencing disestablishment of the horse from its unique position as the main agent by which man and his possessions have been carried and drawn all over the face of the earth will go, it is difficult to say at present.

A hundred years later it can be said that there are very few places in the world where the horse has not been superseded by mechanical transport. It is therefore hard to comprehend today the enormous importance of the horse in the development of almost all the great civilizations of the world. Perhaps the most poignant relic of 'the age of the horse' can be found in the term 'horsepower' which is still used to calculate the power of an engine. It is a measure of the drawing-power of a horse and according to the International System of Units one horsepower (hp) is equivalent to 746 watts and one metric horsepower to 736 watts. The term dates from the beginning of the nineteenth century when horses were in widespread use as providers of power to engines and machines, for example for grinding, spinning, and furnace blowing. According to Tann (1983) 1 hp could crush 32 bushels of malt per hour in a brewery. However, steam soon took over the production of power in industry and transport, and so the long and often cruel servitude of the horse began to diminish.

Fig. 0.4 A horse being used to pump water from a mine on a copper halfpenny token of Low Hall colliery, Cumberland, England 1797. (Photo reproduced by courtesy of the Trustees of the British Museum.)

PART I

WILD HORSES AND ASSES

1 The family Equidae

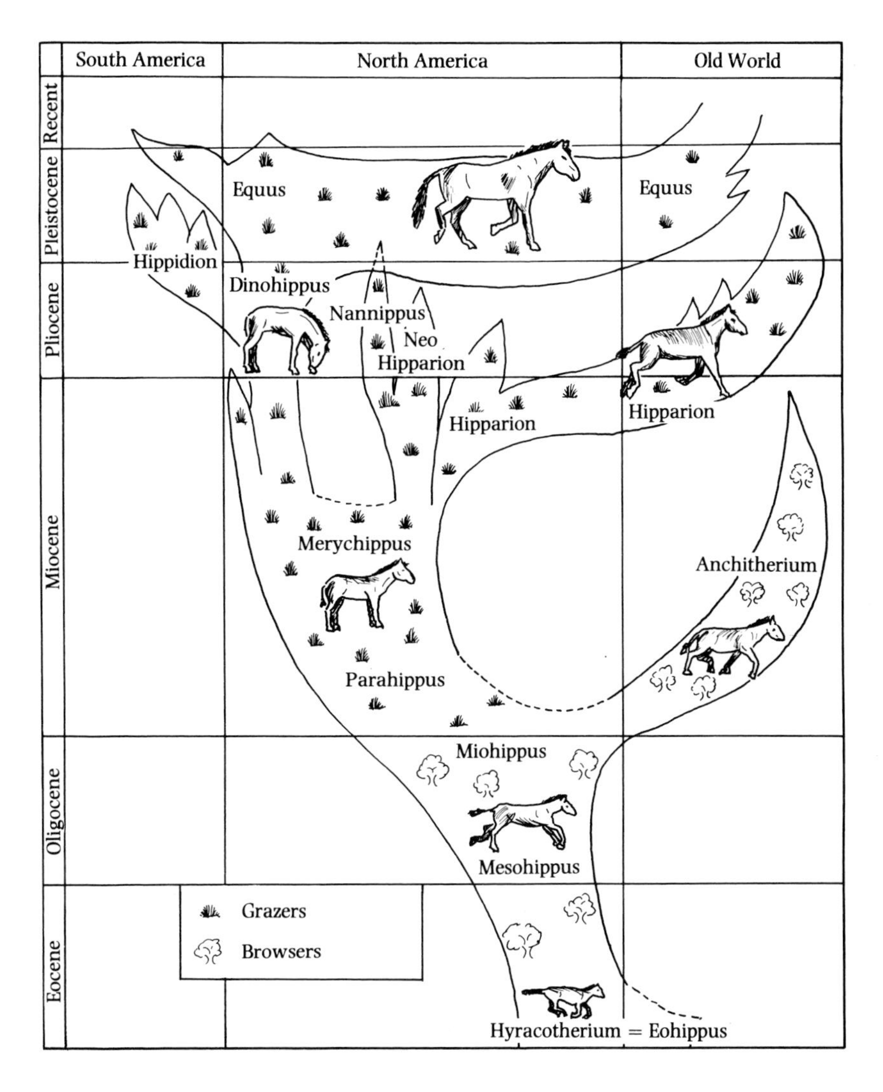

Fig. 1.1 The lineages of the horse family. After Simpson (1961).

There are three main orders of ungulates: the Perissodactyla (comprising tapirs, rhinoceroses, and equids); the Artiodactyla (camelids, pigs, cattle, antelopes, goats, and sheep); and the Proboscidea (elephants). The Equidae, that is the horses, asses, and zebras, are the most highly evolved of the Perissodactyla. The earliest equid in the fossil record dates back sixty-five million years, to the Eocene period, when a small browsing mammal called *Hyracotherium* (*Eohippus*) inhabited the forest regions of North America. *Eohippus* had four toes on the front feet and three on the hind feet. Figure 1.1 shows how the ancestral equids evolved into grazers and dispersed over nearly all the grassland areas of the world until, in the Pleistocene period, they inhabited South as well as North America, Asia, Europe, and Africa.

All ungulates are herbivorous hoofed mammals, and are to a greater or lesser extent digitigrade (or more strictly unguligrade), that is they walk on their toes and their toe bones are increased in length which enables them to run fast to escape predators. Artiodactyla are even-toed, and the weight of the body is borne on the third and fourth toes. In all Perissodactyls the weight of the body is borne on the third toes. Tapirs are characterized by having four toes on the front feet and three on the hind feet: rhinoceroses have three toes on all their feet. In the Equidae, which have the greatest endurance for high speed, toes one and five are missing and two and four are reduced to vestigial 'splint bones', so they appear to have only one toe, the third, on each foot (Fig. 1.2).

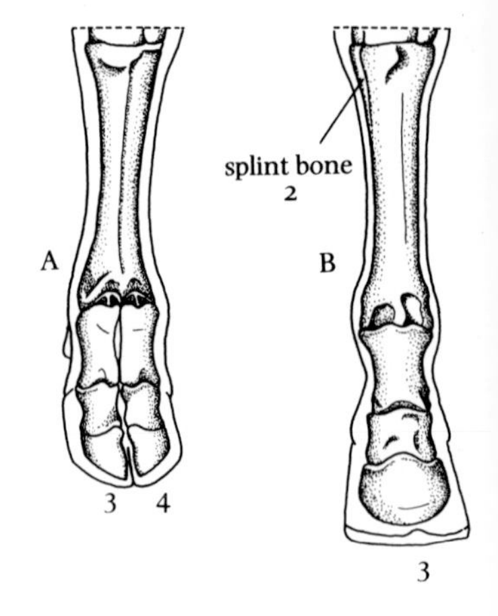

Fig. 1.2 Bones of the left fore-foot of A, ox and B, horse.

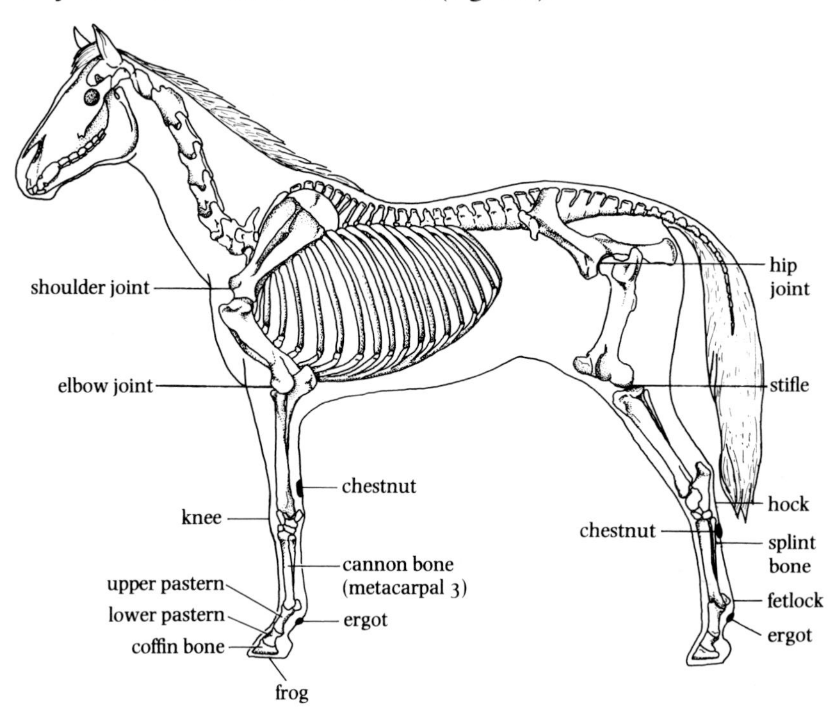

Fig. 1.3 The principal bones and limb joints in the skeleton of the horse.

Besides the splint bones, the 'chestnuts' and 'ergots' are also usually considered to be vestiges of the ancestral five-toed limb. The chestnut is an elongated horny outgrowth of the skin which is normally found on all four limbs of the horse, but only on the forelimbs of the asses and zebras. In the forelimb the chestnut is placed towards the back, above the carpal joint ('knee'), while in the hind limb (in the horse only) the chestnut is below the tarsal joint ('hock'), as shown in Figures 1.3 & 1.4. The ergots are also horny but are smaller than the chestnuts and round in shape: they are found on

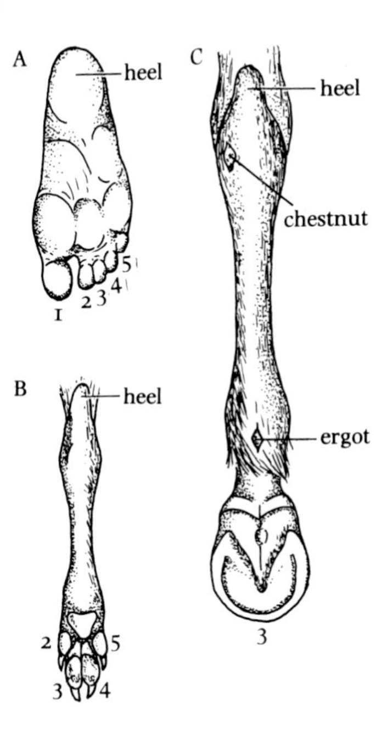

Fig. 1.4 The foot of A: human; B: dog; C: horse to show the corresponding digits. After Flower (1891).

the lower surface at the back of each fetlock joint (Fig. 1.4) in all equids.

At the end of the last century the origins of the apparently functionless chestnuts and ergots were a source of great interest to biologists who were attempting to understand and interpret mammalian evolution. Flower (1891) convincingly argued that the chestnuts had a glandular origin, and that the ergots were vestiges of the trilobed foot pad of those animals that walk on the palms of their hands and feet (plantigrade), such as the dog (Fig. 1.4). At the present day, this explanation is still supported by veterinary anatomists, although it is not upheld in the modern edition of Sisson & Grossman's classic textbook where the chestnuts are considered to be vestiges of the first digit and the ergots vestiges of the second and fourth digits found in the extinct ancestors of the equids (Getty 1975).

Reproduction

Zebras, horses, and asses (which include the African wild ass, domestic donkey, and onagers) are all classified as separate species within one genus, *Equus*, as tabulated in the Appendix. Their taxonomic separation at the species level is supported not only by differences in superficial appearance like coat colour, length of ears, and shape of tail but also by physiological and chromosome differences. The gestation periods and chromosome numbers for the equids are:

LENGTH OF GESTATION	DIPLOID CHROMOSOME NUMBER*
domestic horse 335–346 days	2n = 64
Przewalski horse 333–345 days	2n = 66
African ass and domestic donkey 365–370 days	2n = 62
Asiatic asses 365–368 days	2n = 54–56
zebras more than 365 days	2n = 32–46

Observations on the feral† horses of the Great Basin in North America have shown that horses, away from human control, are relatively slow breeders (Berger 1986). The average reproductive lifespan of both males and females is from five to seventeen years but the mares will not foal each year. The long gestation period of nearly a year ensures that the foal will be born with advanced physical co-ordination so that it has the best chance of escape from predators and is able to keep pace with the moving band. Even so, the mortality of foals is high and disruption of the band or disturbance of the home range may prevent the mares reproducing again for a number of years. Stallions will not mate unless they are able to hold together a harem and any disruption or harassment may turn them into non-breeding lone males.

* There are two sets of homologous chromosomes in the nucleus of each cell of the body, so the usual convention when counting the chromosomes is to cite this double or diploid number. From several sources including Hsu & Benirschke 1967–71.

† Feral animals are defined as those that live in a self-sustained population after a history of domestication.

All horses, asses, and zebras can interbreed but the hybrid offspring are only very occasionally fertile (Gray 1975). Although the wild asses of Asia, the onagers and Kiang, look rather like the African asses, they belong to separate species (*Equus hemionus* and *Equus kiang*: see the Appendix). If an onager is mated with an African wild ass or with a domestic donkey it will produce offspring that are infertile like the hybrid (mule) between a donkey and a horse (see Chapter 3).

Teeth and food

All equids are grazers with a digestive system that allows them to feed on large quantities of low protein fodder. They are therefore able to live in semi-desert and steppe regions, where ruminants cannot thrive, but, because they need such large amounts of fodder, they must be able to range over very large areas and remain in relatively small numbers. Large herds will only occur in areas of exceptionally high quality mixed grasses, but even here the apparent 'herd' will consist of combined family groups or bands of a stallion with a few mares and their offspring.

The grasses that form the staple diet of equids are tough and hard to digest as well as containing particles of silica, called phytoliths, in their cell walls. Evolution follows parallel lines in plants and animals so that phytoliths are present in the cell walls of grasses as a protection against grazing animals, while equids have evolved high crowned teeth to enable them to combat the phytoliths and graze efficiently. Being opportunists, like most mammals, however, horses and particularly asses will browse on bushes, scrub vegetation, and bark when available.

All mammals have two sets of teeth; the milk, or deciduous, and the permanent. In the equids both sets have evolved into a battery of grinding teeth (premolars and molars) with patterns of growth and wear which ensure that the permanent teeth are not worn away completely during the animal's natural lifespan (although this can happen with domestic animals that have an abnormally long life or inappropriate food). The crowns of both sets grow for a number of years and the roots of the permanent teeth do not develop until the animal is at least five years old. After this the long crowns continue to erupt slowly through the jaws, as they are worn down, until in extreme old age the equid will have almost no crown left and the teeth are held in place by the roots, which have increased in size by the deposition of cementum (Fig 1.5A–C).

In front of the grinding teeth, and separated from them by a gap known as the diastema (called the bars in the lower jaw) there are the canine teeth, which are quite large in the male but very small or

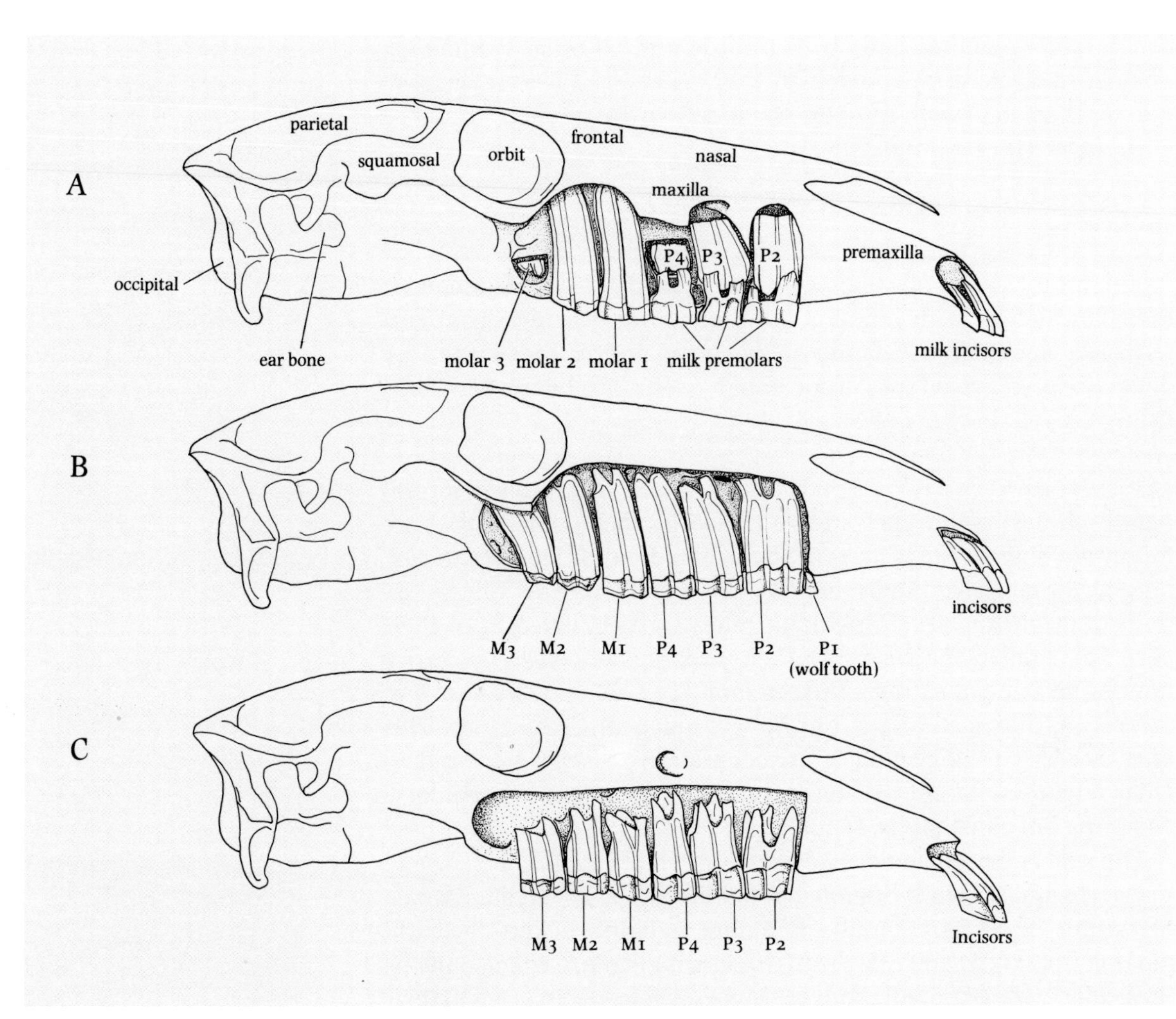

Fig. 1.5 Right side of the skull of a horse with the maxillary bone cut away to show the development of the teeth.
A: At two years old the milk premolars are worn but still in place; the first and second permanent molars have erupted but have no roots; the third molar is half-formed.
B: At five years old all the permanent teeth have erupted. They are still very high crowned and the roots are short and open.
C: At nineteen years the horse is growing old; the crowns of the teeth are much worn and the roots are closed.

even absent in the female, and the incisors. The incisors are used for clutching grass and drawing it into the mouth and, with the canines, for fighting, grooming, and communicating with other members of the herd. The diastema, which is present in all grazers, lengthens the facial region of the head so that the animal can feed with its head down, but its eyes are still at a height where it can watch for predators.

The diastema is of course invaluable to human riders because this is where the bit (mouthpiece of the bridle) is placed, but this is only one of very many features that have made the horse and the ass the ideal means of human transport for thousands of years. More important for the traveller over long distances is the arrangement of the digestive system which, in the equids, does not have a divided stomach with a rumen but instead has a highly developed caecum and large intestine in which the breakdown of cellulose takes place. This means that the equids do not require a period of rest so that they can ruminate after feeding, unlike cattle and all other ruminants. Furthermore, because equids will feed on many different kinds of grasses there are few parts of the world where the traveller would be without feed for his animals.

Behaviour

As important as the morphology and physiology of the Equidae for their association with humans are their natural behavioural patterns. All equids live in family or bachelor groups but within the different species two types of social behaviour can be distinguished (Klingel 1974; Berger 1986):

TYPE I BEHAVIOUR is found in the horse (*Equus ferus przewalskii* and *Equus caballus*), in Burchell's zebra (*Equus burchelli*), and in the rare mountain zebra (*Equus zebra*). These equids form non-territorial, long-term, cohesive family bands of one stallion with one to six mares and their young. Sub-adult mares are abducted from their parental bands by bachelor or family stallions who thereby start a new family band or increase their harem. The stallion will guard and protect his mares and foals, but he will not guard a territory. This type of social organization is an adaptation to migration and to environments with unpredictable conditions as well as to a regularly changing but constant food supply.

The wild horse will not defend a territory, but a stallion will defend his mares and within the family band there is a powerful dominance hierarchy. On a migration, the dominant stallion will keep the band moving from the rear while the highest ranking mare will usually be the leader of the band. All the other horses will follow in order of their dominance. The foals follow their mothers in order of age with the youngest first. These behavioural patterns explain how a convoy of horses or a cavalry charge can be held together and kept moving by a human rider taking over the position of the dominant leader whom all the rest will blindly follow. This behaviour also explains the apparently aggressive actions of a free-living stallion who will try to form a band by abducting mares and will then bite, kick, and rear in his efforts to protect them.

Although the long-ranging and high speed mobility of the horse makes it the ideal means of transport for humans, the process of domestication cannot have been easy, which may partially explain why the horse was the last species of livestock to be enfolded into human societies. Early on in the history of early husbandry and before the practice of castration became widespread the management of a group of domestic horses must have required considerable courage and knowledge of animal behaviour. The wild ponies would

Fig. 1.6 Mules and zebras harnessed together in the Transvaal, South Africa. From Tegetmeier & Sutherland (1895).

have been quite small, about the height of a present-day New Forest pony, but they would have been extremely strong and very difficult to 'break in' to harness or for riding.

TYPE 2 BEHAVIOUR is found in the African wild ass (*Equus africanus*) and hence the domestic donkey, (*Equus asinus*), in Grevy's zebra (*Equus grevyi*), and in the Asian wild asses (*Equus hemionus* and *Equus kiang*). In these equids the males are territorial and they will guard their very large territories for long periods, which may exceed ten years. The adults do not form lasting bonds. Mares can range over several male territories and they can be mated by all males within these areas. Males will not mate unless they can hold a territory, and young males will roam in bachelor groups with no access to mares until they have a territory of their own. This type of social organization can be interpreted as an adaptation to environments with specialized but predictable conditions, such as those found in semi-desert and in hot arid regions.

Because the male African ass is territorial it presents problems in management that are different from those of the horse but no less troublesome, as anyone knows who has tried to make a donkey stallion leave its territory against its will. As with the horse the males can be subdued by castration and the females by coercion. Today, the domestic donkey is a very small equid that has been selected to carry the heaviest possible loads for the smallest amount of food, but its progenitor, the African wild ass, is a typical desert animal with long slender legs, a compact body, neat head, and very long ears. The early domestic asses probably closely resembled this form and would have been no shorter in withers height than the horses which were later to be given higher status in the ritual and warfare of the ancient civilizations.

From the studies of their social behaviour there appears to be no intrinsic reasons why the zebras could not have been domesticated. Klingel's observations on the two types of behaviour suggest that Burchell's and the mountain zebra should respond under domestication like the horse (Type 1) and Grevy's zebra should respond like the donkey (Type 2). It is rather strange that no species of zebra was ever used by the peoples of Africa as pack animals: the reasons were presumably cultural rather than biological. In the 1890s Europeans experimented in training Burchell's zebra as a draught animal in the Transvaal (Fig. 1.6). According to reports quoted by Tegetmeier & Sutherland (1895) the enterprise was very successful and it was hoped that zebras could be substituted for mules which had a high death rate from 'horse sickness'. That nothing more seems to have come of this plan was perhaps due to the outbreak of the Boer War.

Equids as hunters' prey at the end of the last ice age

The fossil record reveals that wild equids ranged in vast herds over all the grass steppes across the whole of the northern hemisphere and in South America, until the end of the last ice age, 10 000 years ago. Furthermore, there were wild asses (*Equus africanus*) in North Africa and probably in Arabia; onagers (*Equus hemionus* and *Equus kiang*) in Asia and possibly North America; zebras throughout Africa: while in southern Europe there was a small wild ass-like equid (*Equus hydruntinus*) whose taxonomic position is still little understood (Uerpmann 1987).

All these equids were hunted by the rapidly increasing human populations, but the wild horse was the chief prey and its remains are found, sometimes in great numbers, on many archaeological sites. The herds of equids slowly dwindled and populations that had been contiguous became dispersed and isolated. All species of wild equids in North and South America were extinct by 10 000 years ago. In Europe the wild horse retreated to the steppes of central Asia and by 7000 years ago it was probably a very rare animal in the West.

Explanations that can be given for the decline of the wild horse at the end of the last ice age are over-hunting and harassment by humans combined with the warming of the climate which caused the grasslands to be replaced by widespread forests. Quite why the equids were so peculiarly susceptible to these pressures is not fully understood, but a modern paradigm can be seen in the rapid extermination by over-hunting and loss of habitat of the Syrian onager (*Equus hemionus hemippus*) and the South African quagga (*Equus quagga*) during the last century.

The famous Upper Palaeolithic site of Solutré in east-central France has been known for a hundred years as a site where wild horses must have been slaughtered in very great numbers during the last ice age. Until recently it was assumed that the horses were simply driven to their deaths over the cliff that borders the site. However, a new and detailed study of the fossil remains re-evaluates the topography of the site in relation to the probable migration routes of the horses, the ecology and behaviour of the animals in life as deduced from observations on living equids, and the butchery of the bones (Olsen 1989). The most likely hypothesis, according to Olsen, is that the hunters followed the bands of horses, probably numbering between six and twelve individuals, as they passed through the valley in their seasonal migrations from the winter grazing grounds on the Saône floodplain to their summer grazing in the highlands. The bands would be driven a short distance into a cul-de-sac formed by the southern cliff face where they were briefly corralled and then killed with spears.

This theory is more acceptable than the idea that huge herds of horses were driven to their death over the cliff because it is based on knowledge of the behavioural patterns of horses living in the wild. The site is over one hectare in area and has fossil deposits up to nine metres in depth. Only a very small sample of the vast deposits of animal bones, of which most were from horses, was studied. From this sample Olsen gives a rough estimate of a total of 32 000 to 100 000 horses killed over a period of 20 000 years, which only amounts to a very small number of deaths each year. The most revealing feature of this new appraisal of the horse remains from Solutré is the continuity that it demonstrates for the seasonal migration of the equids and the human hunters who followed them over many thousands of years. That meat was always plentiful is suggested by the very small amount of butchery of the bones, and indeed perhaps only certain parts of the carcass, like the tongue, heart, or liver were eaten.

It can be seen from the observations by Berger (1986) on the ecology and behaviour of the feral horses in Nevada that any disruption of the social group will adversely affect reproduction. The expanding human populations at the beginning of the Holocene, combined with climatic change, would have made such disruption very widespread. Hunting and harassment would alter the structure of the herds and sometimes remove the breeding stallion or mare. Natural grasslands would be depleted, migration routes would be distroyed through forestation, fire, or human settlement, and home ranges would be lost. Berger reported that stallions were remarkably faithful to their annual use of home ranges: 99 per cent of his study animals returned to their low-altitude core area after the summer migration to higher regions. Over a period of a few thousand years at the end of the last ice age, it seems that human activity, combined with the spread of the forests throughout Europe, was sufficient disruption to drive away to the east the populations of wild horses: while in North America all the species of equids were exterminated, along with camelids, antelopes, and the mammoths.

2 Wild horses and asses in historic times

Apart from the small numbers of Przewalski horses that are at present being re-introduced to the wild in Mongolia and in China (Ryder 1988; Sattaur 1991), it is extremely unlikely that there are any populations of wild horses alive today that are not mostly descended from domestic stock. Until the end of the eighteenth century, however, there were two subspecies of the wild horse in Europe and Asia: the tarpan in eastern Europe and the Russian steppes, and the Mongolian wild horse, or Przewalski's horse, in Mongolia (see the Appendix). These two wild equids, which survived in ever-dwindling numbers, were the relics of vast populations of wild horses that were distributed over nearly the whole of Europe, Asia, and North America at the end of the last ice age.

Fig. 2.3 Reconstituted tarpan in Poland. (Photo G. Barker.)

Fig. 2.5 Przewalski's horse, *Equus ferus przewalskii*. (Photo Sally Anne Thompson.)

Fig. 2.10 The Nubian wild ass, *Equus africanus africanus*. From Lydekker (1904).

Fig. 2.11 *Left*: The Somali wild ass, *Equus africanus somaliensis*. *Right*: Th Nubian wild ass, *Equus africanus africanus*. From Sclater (1884).

Wild horses

THE TARPAN, *Equus ferus ferus*

That there were wild horses in many parts of Europe in classical times seems not to be in question: the Greek historian Herodotus wrote of wild white horses in Scythia and they have been described by Strabo and the Roman writers Pliny, and Varro. The question remains, however, of whether these horses were truly wild or whether they were feral herds, perhaps of mixed origin. A detailed description of the wild and feral horses of Europe and central Asia from historical and contemporary accounts was given by Hamilton Smith (1845) and summarized by Lydekker (1912). Hamilton Smith recognized the difficulties of distinguishing the truly wild herds from the feral but he had been informed that:

> It does not appear that the Tahtar or even the Cossack nations have any doubt upon the subject, for they assert that they can distinguish a feral breed from the wild by many tokens; and naming the former *Takja* and *Muzin*, denominate the real wild horse *Tarpan* and *Tarpani*.

This is the first use in English writing of 'tarpan', the Turkoman word for the wild horse, but since the description of Przewalski's horse by Poliakof in 1881, it has only been used to describe the now extinct western group of wild horses. Hamilton Smith gave this description of the wild horses:

> The real Tarpans are not larger than ordinary mules, their colour invariably tan, Isabella [greyish-yellow], or mouse, being all shades of the

2.3

2.5

2.10

2.11

same livery, and only varying in depth by the growth or decrease of a whitish surcoat, longer than the hair, increasing from midsummer and shedding in May; during the cold season it is long, heavy, and soft, lying so close as to feel like a bear's fur, and then is entirely grizzled; in summer much falls away, leaving only a small quantity on the back and loins: the head is small, the forehead greatly arched, the ears far back, either long or short, the eyes small and malignant, the chin and muzzle beset with bristles, the neck rather thin, crested with a thick rugged mane, which, like the tail, is black, as also the pasterns which are long; the hooves are narrow, high, and rather pointed; the tail descending only to the hocks, is furnished with coarse and rather curly or wavy hairs close up to the crupper; the croup as high as the withers: the voice of the Tarpan is loud, and shriller than that of a domestic horse; and their action, standing, and general appearance, resembles somewhat that of vicious mules ...

The genuine wild species is migratory, proceeding northward in summer to a considerable distance, and returning early in the autumn ...

The Tarpany form herds of several hundred, subdivided into smaller troops, each headed by a stallion; they are not found unmixed, excepting towards the borders of China; they prefer wide, open, elevated steppes, and always proceed in lines or files, usually with the head to windward, moving slowly forward while grazing, – the stallions leading and occasionally going round their own troop; young stallions are often at some distance, and single, because they are expelled by the older until they can form a troop of young mares of their own.

During the Middle Ages the wild horses of eastern Europe were hunted as game animals and their meat was highly esteemed by the aristocracy. In 1409 King Wladislaw Jagiello arranged a great chase of wild horses in the forest of Bialowieza in Poland for his cousin, Witold of Lithuania (Antonius 1937). Tarpans survived in the forest regions of Poland, along with bison (*Bison bonasus* which are still there today) and the aurochs (*Bos primigenius*, extinct by 1627) until between 1810 and 1820 when the last ponies from the final herd were captured and given to the local farmers. The present day native ponies of Poland, known as Koniks, bear a close resemblance to the historical descriptions of the tarpan and it is probable that they are partly descended from the last wild individuals. In particular, the Konik shares with the tarpan the characteristic of a whitening of the coat in winter (Bökönyi 1974*a*).

The only engraving of a tarpan known to have been drawn from life appears in the book of G. S. Gmelin's travels in Russia (1770). The animal was a year old filly and shows only one distinctive feature, the short upstanding mane which is a characteristic of all wild equids (Figs. 2.1 & 2.2). Gmelin wrote the earliest scientific description of the tarpan and this led to Antonius naming it *Equus gmelini*. However, Boddaert (1785), also on the basis of Gmelin's description,

Fig. 2.12 Domestic asses in Ethiopia. (Photo author.)

Fig. 2.15 Mustangs in the Nevada desert. (Photo Bob Langrish.)

Fig. 2.18 Feral ass of Sokotra. From Forbes (1903).

Fig. 2.1 The tarpan, *Equus ferus ferus*, as seen by Gmelin (1770).

Fig. 2.2 The tarpan of Hamilton Smith (1845).

had named the tarpan *Equus ferus*, and as this is therefore the nominate, or first description of the wild horse, the correct Latin name for the tarpan is *Equus ferus ferus* (Antonius 1937, Groves 1974).

Gmelin described the Russian tarpan as a small, clumsy headed, mouse-coloured horse, with a short wavy mane and the fronts of the legs black from the knees and hocks downwards. In some examples the ears were short and horse-like, but in others longer and more ass-like. The tail was in some cases bushy and in others scantily haired, but always shorter than in domesticated horses. However, it is probable that Gmelin was describing equids that were mostly hybrids, and that even in his day the true wild horse was a rare animal (Lydekker 1912). Today, the only physical remains of the tarpan are one complete skull and skeleton in the Zoological Institute, St. Petersburg, and one skull in the Institute for Evolution History, Moscow (Bökönyi 1974*a*).

There have been two attempts to 'breed back' the tarpan. In Germany before World War II in the zoos at Munich and Berlin the Heck brothers crossed Przewalski horses with mouse-dun mares of various domestic breeds, while in the Bialowieza Forest in Poland selective breeding of the local Konik and other ponies produced a breed that is considered to look very like the tarpan and which is flourishing today (Groves 1974; Figs. 2.3 & 2.4).

PRZEWALSKI'S HORSE, *Equus ferus przewalskii*

At the time of its first description by the Russian explorer Przewalski in 1876, this wild horse was to be found in a number of localities in the Gobi desert and the steppes of Mongolia. It was named *Equus przewalskii* by Poliakof in 1881, but as it is now considered to be

Fig. 2.4 The Konik. (Photo Sally Anne Thompson.)

conspecific with the tarpan its correct name is *Equus ferus przewalskii* (Groves 1974; Fig. 2.5).

The story of the travels of Przewalski in the 1880s and his account of the wild horse which were the incentive for the expeditions sent, under the patronage of the Duke of Bedford (between 1901 and 1904) to capture specimens from Mongolia, have been told repeatedly and are documented in detail in the two books devoted to the horse by Mohr (1971) and Bökönyi (1974*a*). What is perhaps more interesting today is to examine the arguments for and against the acceptance of Przewalski's horse as a new species of equid when the live animals first reached western Europe.

In 1891 Sir William Flower, first Director of the British Museum (Natural History), and according to Ewart (1904), 'the greatest English authority on the structure and classification of the Equidae during the latter part of the nineteenth century', described the newly discovered equid as follows:

> Much interest, not yet thoroughly satisfied, has been excited among zoologists by the announcement (in 1881) by M. Poliakof of the discovery by the late distinguished Russian explorer, Prjevalsky, of a distinct species of wild horse. One specimen, unfortunately, only was obtained, while searching for wild camels in the sandy desert of central Asia near Zaisan. It is described as being so intermediate in character between the equine and asinine group of *Equidae* that it completely breaks down the generic distinction which some zoologists have thought fit to establish between them. It has callosities on all four limbs, as in the horse, but only the lower half of the tail is covered with long hairs, as in the ass. The general colour is dun, with a yellowish tinge on the back, becoming lighter towards the flanks and almost white under the belly, and there is no dorsal stripe. The mane is dark brown, short, and erect, and there is no forelock. The hair is

long and wavy on the head, cheeks, and jaws. The skull and the hoofs are described as being more like those of the horse than the ass.
Until more specimens are obtained it is difficult to form a definite opinion as to the validity of this species, or to resist the suspicion that it may not be an accidental hybrid between the Kiang and the horse.

Cossar Ewart, Professor of Natural History at Edinburgh University, was much intrigued by the new equid and he wrote (1904):

As far as I can gather, it is generally believed in England that Prjevalsky's horse is a hybrid – a cross between a pony and a Kiang. Beddard, however, admits it may be a distinct type. He says: 'This animal has been believed to be mule between the wild ass and a feral horse; but if a distinct form – and probability seems to urge that view – it is interesting as breaking down the distinctions between horses and asses.'
It must be admitted that in its mane and tail Prjevalsky's horse is strongly suggestive of a hybrid, but in the short mane and mule-like tail we may very well have persistance of ancestral characters.

Later in his career Ewart was responsible for some hare-brained ideas about the origins of domestic horses, which have unfortunately remained in the literature, but in his understanding of the Przewalski horse Ewart was entirely accurate, and he took the correct action to establish the validity of the species. For, as he wrote:

Though failing to understand why so many zoologists persisted in considering the horse of the Great Gobi Desert to be a mule I decided to breed a number of Kiang-horse hybrids.

As expected the offspring bore little resemblance to Przewalski's horse. Ewart then continued with his breeding experiments in order to refute the suggestions that the horse was merely a feral Mongolian pony*, or that it was indistinguishable from an Iceland pony. He crossed a chestnut Mongolian pony with a Connemara stallion and with an imported yellow-dun Iceland stallion and was able to show that the progeny did not look at all like Przewalski's horse.

The wild horses soon began to breed in captivity and their numbers increased. Ewart's claim that the short mane, absence of forelock and rather ass-like tail are 'ancestral characters' is justified in the sense that they are the natural characteristics of the equid family. The long mane which is extended into a forelock and the full tail are features that have evolved with the process of domestication (see Chapter 3).

The first Przewalski horses to be imported into Europe were 'a motley lot' (Mohr 1971) that showed considerable diversity of form, but after a number of generations in captivity and careful management of their breeding the horses became more uniform. Over the years changes have occurred that can only be ascribed to the effects of captivity. The cranial capacity (size of the braincase) has been

* A pony is a horse with a height of less than 14 hands at the shoulder. A hand is equal to four inches or 10.16 cm.

reduced, the teeth have become shorter and the muzzle less wide. The age at which the horses become sexually mature has been lowered from five to two years and white marks sometimes appear in the hair whorl on the forehead. In the summer coat the horses often have a dorsal stripe.

There have been no reliable sightings of Przewalski's horse since 1968 and so it must now be assumed that this equid is extinct in the wild, not because it was hunted but because of competition with domestic stock for pasture and water. It has, however, bred well in captivity and a stud book is held at the Prague Zoo. In 1987 there were 626 Przewalski horses in zoos and parks worldwide (Houpt & Fraser 1988). Attempts are under way to re-introduce the horse into the wild in a special reserve in Mongolia.

Wild asses

Apart from the zebras, the most viable of the wild equids today are the Asiatic asses, or onagers, but they are on the endangered list, as is the one remaining race of African wild ass, the Somali ass.

THE AFRICAN WILD ASSES, *Equus africanus*

It is interesting that the European 'big game hunters' of the nineteenth century, men who must be held accountable for the unnecessary slaughter of millions of animals, showed an unusual restraint about shooting the wild ass. Captain Swayne in his record of exploration and big game shooting, from 1885–93, wrote:

> The wild ass is common in sterile parts of Guban, especially to the east of Berbera. In Ogådén its place is taken by the zebra. It is a fine animal and has striped legs. It can scarcely be considered as fair game to the sportsman.

Even so, the African wild ass, *Equus africanus* is all but extinct today, through a combination of loss of habitat, capture of young animals for zoos (with the inevitable slaughter of breeding mares) and interbreeding with domestic and feral donkeys.

At the end of the last century a number of travellers, naturalists, and taxonomists described the extinct and living races of African ass. In 1909 Pocock, who did not accept that the domestic donkey could be descended from one of the living races of wild ass, attempted to summarize the characteristics that distinguished the subspecies as he saw them:

A distinct and large black or brown patch at the base of the ear as well as at the tip; legs the same tint as the body or approximately so (domestic and possibly [ancestral] wild forms).......................... Subspecies *asinus* L.

No large and distinct dark patch at the base of the ear; legs usually, at all

Fig. 2.6 Representation of the extinct Algerian wild ass, *Equus africanus 'atlanticus'* from a rock engraving at Enfouss. From Werth (1930).

Fig. 2.7 Drawing of the extinct Algerian wild ass made by Monod (1933) from a Roman mosaic at Hippo Regius (Bône).

events, markedly lighter than the body and striped or unstriped.
Legs unmarked except for patches on the fetlock ....
Subspecies *africanus* Fitzinger, 1857. [Nubian wild ass]
Legs boldly striped.
Spinal and shoulder-stripes well developed ....
Subspecies *taeniopus* Heuglin, 1861.
Spinal and shoulder-stripes obsolete or nearly so ....
Subspecies *somaliensis* Noack, 1884. [Somali wild ass]

This summary is accurate except that Pocock failed to describe the extinct race from Algeria and the Atlas mountains, *Equus africanus 'atlanticus'* Thomas, 1884 (the name '*atlanticus*' is invalid because it was earlier used for a zebra). This wild ass has been recorded from prehistoric rock art at Enfouss and is represented on a Roman mosaic from Hippo Regius, the modern Bône, on the Algerian coast (Figs. 2.6 & 2.7). It had a well-developed shoulder stripe and strongly striped legs, like *E. africanus taeniopus* from the Red Sea coast, which was first described by the traveller Heuglin (1861; Fig. 2.8). It is not known, however, whether the asses that Heuglin saw were a really wild population because herds of feral and cross-bred donkeys are common all over the Sahara and no distinctive asses conforming to Heuglin's description have been seen during the present century. Another population of asses that was described as *Equus asinus dianae* Dollman, 1935, was also most probably a feral group. Feral asses differ from the wild Nubian and Somali asses in being smaller and more varied in colour. A very interesting feral population, of ancient origin, was described in 1903 from the island of Socotra, off the Horn of Africa, by Forbes (see p. 40; Fig. 2.9).

Fig. 2.9 Distribution of the extinct and living wild asses of Africa and Asia. After Groves (1974).

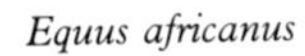

Equus africanus

1. *africanus*
2. *somaliensis*
3. *'atlanticus'*

? Fossil records of *E. africanus*?

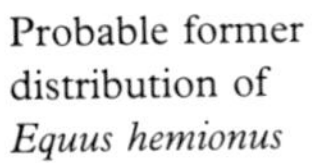

Probable former distribution of *Equus hemionus*

4. *hemious*
5. *luteus*
6. *kulan*
7. *onager*
8. *khur*
9. *hemippus*

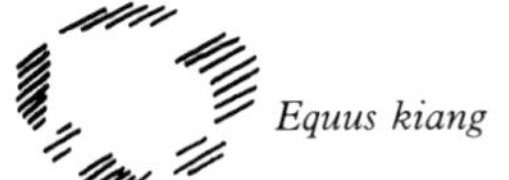

Equus kiang

Fig. 2.8 The wild ass of Heuglin (1861), *Equus africanus taeniopus*.

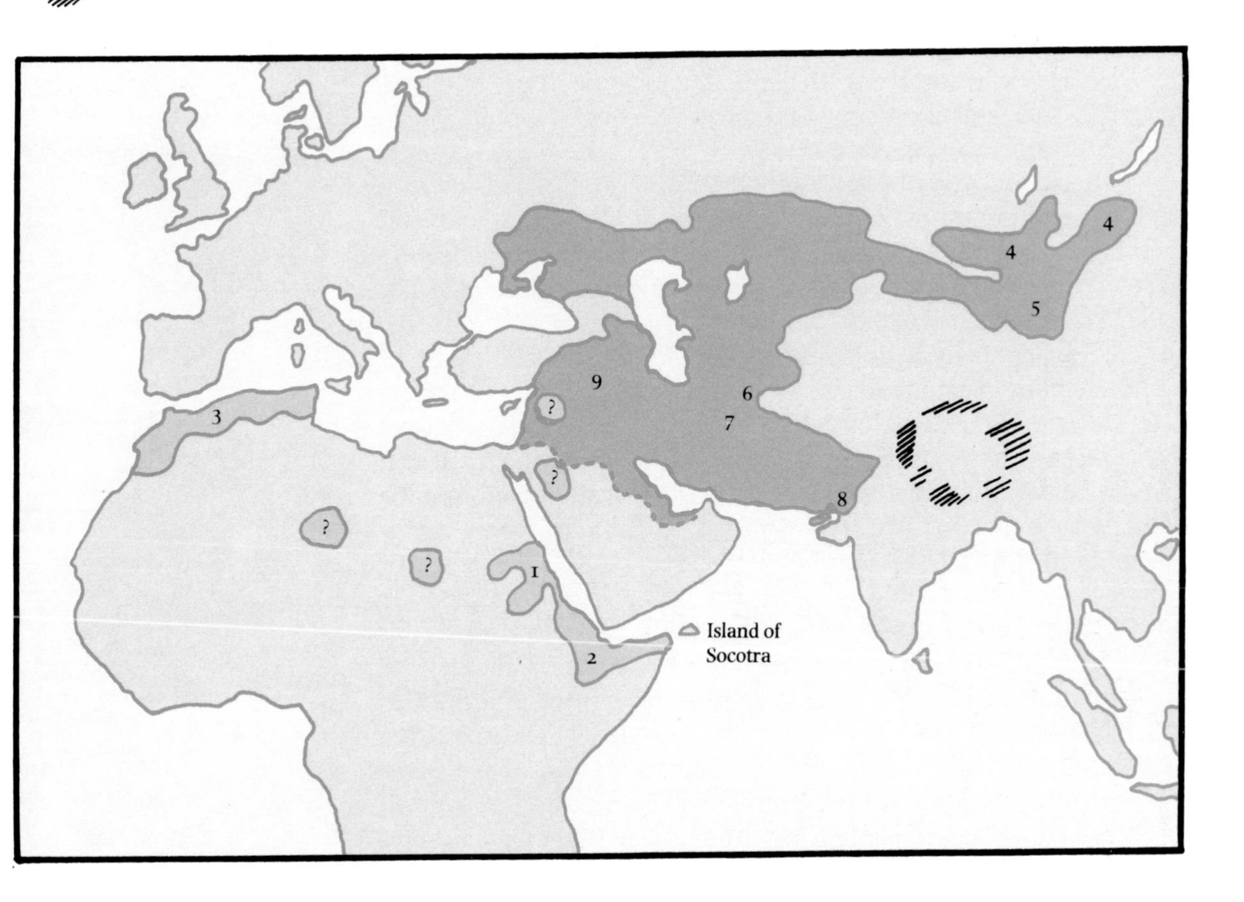

The Nubian wild ass (*E. africanus africanus*) is probably extinct in the wild today; in the last century it was found from the eastern Sudan to the Red Sea, but it probably had a much wider distribution in ancient times (Fig. 2.9). The Nubian wild ass is depicted as a hunted animal in ancient Egyptian art and it is this race, with a marked shoulder stripe but no leg stripes, that is always shown as a domesticated beast of burden in the ancient world (Fig. 2.10).

The Somali wild ass (*E. africanus somaliensis* = *Asinus somalicus* Sclater, 1884) is the only race to be found today in the wild. It lives, as an endangered species, in Somalia and in the drought-stricken and war-torn Danakil region of Ethiopia (Fig. 2.9). The Somali ass is distinguished by strong banding around the legs while it may or may not have a shoulder stripe. It is a long-legged, beautiful and powerful equid which bears little resemblance to the small domestic donkey of northern regions (Fig. 2.11).

Although the domestic asses of the ancient Egyptians appear to have been always of the Nubian type, as they were never depicted with stripes on the legs, the present day donkeys of Ethiopia must be of mixed origins for they often have the strong shoulder stripe of the Nubian, combined with the leg stripes of the Somali ass (Fig. 2.12). In support of this Groves (1974) quotes a description by the Italian naturalist Ziccardi of how the local people of the Danakil left she-donkeys beside water-holes at night, hoping that they would be mated by wild Somali asses, and produce improved offspring.

The African wild asses cannot survive in the sand-dune wastes of the Sahara, but they thrive in any dry stony area where there is scrub vegetation on which they can graze, and a supply of water within two or three days reach. The drastic decline in their numbers is partly due to military activity in Somalia, Sudan, and Ethiopia, but also and perhaps more importantly, as with Przewalski's horse, to severe competition with domestic livestock for pasture and water.

Recent finds of equid remains from archaeological sites in Arabia are providing increasing evidence for the presence of a wild ass, belonging to the species *Equus africanus*, in the prehistoric period. It cannot be precluded therefore that the earliest domestication of the donkey took place in western Asia rather than in North Africa (Uerpmann 1991).

THE ASIATIC WILD ASSES, *Equus hemionus* AND *Equus kiang*

In the ancient world there were wild asses or onagers all over the lowland deserts of western and central Asia from Anatolia to Mongolia, while the Kiang, which is now considered to be a separate species (Groves 1974), inhabited the Tibetan plateau above 3000 metres, where it is still to be found in small numbers. As with all other wild ungulates the Asiatic asses have been killed for sport and

Fig. 2.13 Iranian onagers, *Equus hemionus onager*, with oryx antelopes, in Israel. (Photo Dr. Colin P. Groves/Eve Bygrave.)

Fig. 2.14 The Indian onager or khur, *Equus hemionus khur*. From Lydekker (1904).

for their skins, and their habitats have been eroded away so that today the most western subspecies, the Syrian onager, is extinct, and the others cling on in four small separated regions (Fig. 2.9 and Appendix).

Unlike the African wild ass, which is the progenitor of the domestic donkey, the Asiatic asses have no domesticated descendants. It has often been contended that onagers were domesticated in the ancient world but this appears from the biological evidence to be unlikely. If the domestic asses of Asia had been derived from the onager they would produce fertile offspring when mated with the wild species, but this is not so (Gray 1971). All domestic asses that have been mated with wild species produce fertile offspring with the African ass, and infertile offspring with the onager and Kiang.

The onager is also known as the half-ass and sometimes as the stilt-legged ass because its lower limb bones (cannon bones and pasterns) are relatively long. Besides being longer-legged than the African ass or domestic donkey the onager has no shoulder– or leg-stripes but it usually has a dark stripe along the length of the back. The different subspecies of onager and the Kiang have coats that vary in colour from sandy yellow to reddish to pale buff, with white underparts (Figs. 2.13 & 2.14).

In ancient Mesopotamia the Syrian onager was hunted for sport and probably for its hide. Captured animals were also bred with domestic donkeys and later with horses to produce mules (p. 85). This most western onager (*Equus hemionus hemippus*), a very small wild ass that hardly reached a metre at the withers, became extinct early in the present century with the last individual dying in the Vienna zoo in 1927. It was the wild ass of the Bible, with the Hebrew name *pere*, and is described in a number of evocative passages, as in the Book of Job (39, 6–8):

Whose house I have made the wilderness, and the barren land his dwellings. He scorneth the multitude of the city, neither regardeth he the crying of the driver. The range of the mountains is his pasture, and he searcheth after every green thing.

Tristram (1889) claimed that the Syrian wild ass was, like all other wild equids, migratory. He wrote that it travelled north and south, according to the season, in large herds:

The Asiatic [asses] (*Asinus hemippus*) proceed in summer as far north as Armenia, marking their course by grazing the herbage very closely on their march. In winter they descend as far as the shores of the Persian Gulf.

Like the Syrian onager the Indian wild ass or khur, which is today an endangered race, was sometimes interbred with domestic donkeys or horses to produce mules. The khur inhabits the desert of the Little Rann of Kutch in northern Gujarat (Smielowski & Raval 1988). It has been described by Tegetmeier & Sutherland (1895) in one passage as wild and totally untameable, and in another as being as docile as a tame donkey. Presumably, as with the taming of all mammals, its response to humans depended on the way it was treated.

Feral* horses and asses

There are populations of feral horses and asses throughout the habitable world and it is reasonable to assume that ever since the first equids were domesticated, five thousand years ago, there must have been an interchange between the wild and the tame. This is because horses and asses, being extremely hardy as well as feeders on ubiquitous grasses, can survive in many different environments and, once they are away from human control, they can move fast over long distances to reach new areas. In addition, the natural instincts of domestic mares will ensure that they will readily join a stallion-led band of feral horses, or a feral stallion will 'kidnap' domestic mares if he can.

Berger (1986), who carried out a detailed demographic and behavioural study of the feral horses of the Great Basin desert in Nevada, quotes a figure of two million for the number of feral horses, or mustangs, in North America during the nineteenth century. Today these numbers have been reduced to 45 000 but herds of feral horses are still found in many of the western United States (Fig. 2.15).

In Europe the best known feral horses are the white ponies living in the Camargue marshlands of the south of France, and the New Forest and Exmoor ponies in England. These, as well as the Iceland pony and other populations of moorland ponies are probably of very

*For a definition of feral see p. 19.

Fig. 2.16 *Below*: Camargue ponies. (Photo Sally Anne Thompson.)

Fig. 2.17 *Right*: Exmoor ponies.

ancient origin and breed true to type (Fig. 2.16). The Camargue ponies have particularly wide hooves in adaptation to the marshy land that they live on, while the Exmoor pony has the dun colour and 'mealy' muzzle of the wild horse, *Equus ferus* (Fig. 2.17). The general similarity of the Exmoor pony to Pleistocene cave paintings from France and Spain of wild horse, as well as to illustrations of the Tarpan and to Przewalski's horse, have given rise to the modern myth that this pony is indeed a relic of the wild horse, *Equus ferus* (Speed & Speed 1977). However, earlier writers have given details of the Exmoor's history and the many efforts that were made in the

nineteenth century and earlier to improve the ponies by crossing with, amongst other breeds, Dongola Arabs (Gilbey 1903).

Despite their mixed ancestry and the considerable extent to which they have been managed, the feral and 'native' ponies of Britain and other countries in Europe do comprise long-established populations with particular adaptations that have evolved in response to the harsh environments in which the ponies live.

The largest populations of feral donkeys, numbering many thousands, are to be found in the western United States of America where, like ordinary domestic donkeys, they are called burros. Donkeys were first introduced into the Americas by the Spanish in the seventeenth century and probably soon afterwards became feral. But their numbers did not increase greatly until the nineteenth century, because until the coming of the railways, the donkey was too valuable to both Europeans and native Americans to allow it to escape into the wild (Lever 1985). Through overgrazing and competition for scarce water supplies the large numbers of burros are today a serious threat to the fragile environment and fauna of the Nevada desert.

Feral donkeys are also responsible for soil erosion and for degrading large areas of pasture in the interior and northern parts of Australia. Donkeys were taken there in large numbers during the nineteenth century for use as pack animals but, as in America, the coming of the railways made them redundant so that by the 1950s their numbers were estimated to be somewhere around 150 000. Since that time the donkeys have been extensively culled but there are still many thousands living wild.

Both horses and donkeys live as feral populations on the Galapagos and other Oceanic Islands (Lever 1985) but the only record of anciently established feral equids on an island are the asses on Socotra, which lies 193 kilometres east of the tip of the Horn of Africa. These asses were seen and described by Ogilvie-Grant and Forbes on their expedition to Socotra in 1898–9. The asses were living wild and were very small (about one metre withers' height), mouse-coloured, with a dark shoulder and back stripe and faint bars around white legs (Fig. 2.18). Their uniformity of size and colour pattern indicated that the population had been established on the island for a very long time and they did not interbreed with the domestic donkeys. Forbes (1903) considered that the Socotran asses more closely resembled the Nubian wild ass (*Equus africanus nubianus*) rather than the Somali ass, which would be geographically closer, and he speculated that they had been taken to the island as domestic donkeys in ancient Egyptian or Roman times. If this population of feral asses still survives on Socotra it could be an invaluable genetic relic of the extinct Nubian wild ass.

PART II

EQUIDS THAT ARE WITHOUT PRIDE OF ANCESTRY OR HOPE OF POSTERITY

3 Hybrids and the breeding of mules

3.1

3.2

Hybrid vigour

All the different species of equids, the zebras, African asses (including the donkey), Asiatic asses, and the horse can interbreed (Figs. 3.1, 3.2, 3.3). The offspring, although they are almost always sterile, have the very great benefit for humans in exhibiting heterosis or hybrid vigour. This means that the hybrid is likely to be larger in body size, have greater endurance, and survive better on poor food than either of its parents.

Hybrid vigour is believed to result from the increase of heterozygosity* caused by the interbreeding of two genetically different individuals. The sterility of the hybrids does not mean that they lack either internal or external reproductive organs and, to make it tractable, it is as necessary to geld a male mule as a stallion horse. The male hybrid is unable to produce spermatozoa because the two kinds of chromosomes of its parents are unequally matched (see p. 19 for the chromosome numbers of the equids). The female hybrid, although she may come into oestrus and occasionally may accept the male, will only very rarely produce a viable offspring. Tegetmeier & Sutherland (1895) reported that it was not uncommon for a female mule to foster and even suckle the foals of other mares, even though she had never had a foal herself.

Fig. 3.1 *Right*, hybrid between a donkey and an onager; *left*, hybrid between a donkey and a zebra. From Gray (1850), where wrongly annotated.

Fig. 3.2 Hybrids between an onager and a wild ass *(Equus africanus)*. From Antonius (1944).

* A heterozygote is an organism that will not breed true for a particular character because it has received different alleles (forms) of a gene from each of its parents.

Fig. 3.3 Hybrid between a donkey and a zebra in Zimbabwe. (Photo author.)

The first 'mules'

The first people to breed hybrid equids were probably the Sumerians and their neighbours in western Asia at the beginning of the third millennium BC. The hybrids were crosses between the wild onager (*Equus hemionus*) and the early domestic donkey (*Equus asinus*) which was still a rare and valuable pack animal. The wild onager was a common animal of the local countryside, but it was difficult to tame and did not breed well in captivity, while the donkey, descended from the African or Arabian wild ass (*Equus africanus*) was much easier to habituate and control.

The present view on the osteological evidence suggests that the African wild ass did inhabit western Asia at this time (Uerpmann 1991), but it was probably confined to the Arabian peninsula because its remains are not found amongst the spoil of hunters on archaeological sites in the Fertile Crescent, whereas bones of onager are often found in large numbers (see for example Bökönyi 1986). Although it is difficult to distinguish between the bones and teeth of the true ass (wild or domestic) and those of the onager, there are a few key elements that allow discrimination. The most important evidence for the crossing of domestic donkeys with wild onagers, however, comes from the written sources in the form of clay tablets inscribed in cuneiform, which have been excavated in considerable numbers from Sumerian sites in Mesopotamia. There are a number of different cuneiform signs which have been transcribed as meaning wild, domestic, and hybrid equids. The Sumerian texts indicate that donkeys and donkey-onager crosses were commonly bred in the earlier period, that is around 2800 BC, but that later in the third millennium BC the onager ceased to be used and mule-breeding between donkeys and horses became the common practice (Zarins 1976, 1978; Maekawa 1979*a*,*b*; Maekawa & Yildiz 1982; Postgate 1986; see Chapter 7).

Hybrids between onagers and asses, and onagers and ponies were not uncommon until recent years when the onager has become a rare species. Many different hybrids have been bred since the middle of the nineteenth century, including crosses between onagers and donkeys (Antonius 1944; see Fig. 3.1). In the Rann of Kutch in the Rajasthan desert the crossing of donkeys with the wild *Equus hemionus khur* was a common event and was said to produce very fast 'mules'. So the breeding of hybrids between asses and onagers by the Sumerians as a precursor to the production of true mules is not at all improbable.

Fig. 3.4 Poitou mule. From Tegetmeier & Sutherland (1895).

Mules and hinnies

A mule is the progeny of a male donkey (jack or jackass) and a female horse; its Latin name is written as *Equus asinus* × *Equus caballus*. A hinny or jennet is the progeny of a male horse and a female donkey (jenny); its Latin name is written as *Equus caballus* × *Equus asinus*. Mules and hinnies usually have consistent characteristics which

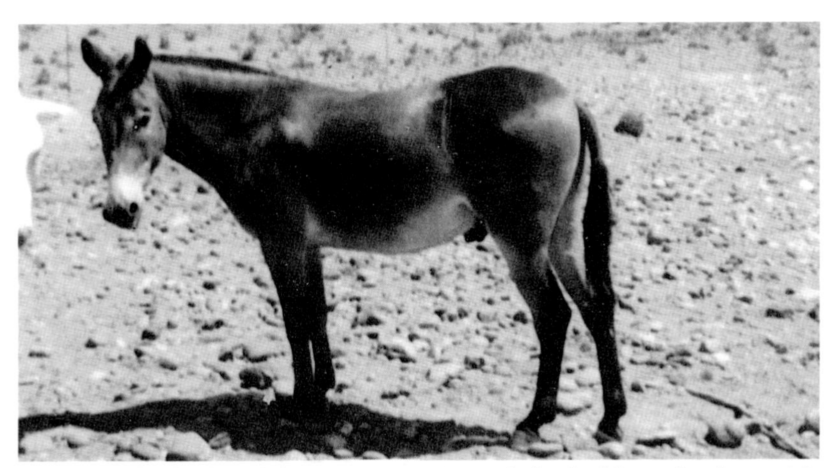
Fig. 3.5 A hinny. (Photo B. Hutchins.)

combine the looks of their parents. A mule looks like a donkey with the body of a horse; its head is heavy, its ears are long, its tail is ass-like, and its legs are fine-boned with small hooves (Savory 1979; Fig. 3.4). A hinny, on the other hand, looks more like a horse with the body of a donkey; its head is lighter, the ears shorter, and the tail fuller (Fig. 3.5). This follows the general rule that the head and tail of the hybrid inherit the characteristics of the sire. The enamel patterns on the biting surfaces of the teeth also combine the characters of donkey and horse. Savory states that the chestnuts are lacking in the mule (presumably on the hind legs, see Fig. 1.3) and that the voice of the mule is, 'a kind of bray, but it is not exactly the voice of the ass'. He attributes the well-known saying about the mule to the nineteenth-century American politician, Ignatius Donnelly: 'the Democratic Party is like a mule – without pride of ancestry or hope of posterity.'

The mule is generally stronger and more robust than the hinny and, at least since the Roman period, this hybrid has been bred more frequently than the hinny, although Tegetmeier & Sutherland (1895) claimed that, in their time, hinnies were much bred in Ireland. Dent (1972) declared that the jennet (hinny) was very popular on Cyprus until recent times as a riding and pack animal. He calculated that there were probably 8370 asses, 1050 mules, and 375 horses on Cyprus in 1967 which he described as a last stronghold of the ass and mule in the Mediterranean region.

The reason that the mule is a taller and more robust animal than the hinny is because the mule's dam, the mare, is larger than its sire, the donkey. When the cross is the other way about the progeny will not be much larger than the donkey mother, for the size of the dam limits the size of the foetus. It is probable that the hybrids bred during Sumerian times were hinnies because a rare imported stallion could produce a whole herd of hybrids in the time it would take one mare to produce a mule (Littauer pers. comm. 1991).

The question of whether female mules can give birth to viable offspring was much debated at the end of the last century. Tegetmeier & Sutherland were adamant that they could not and they went so far as to claim, of a mule in Paris which had produced progeny from both a stallion horse and a jack ass, that:

It is is not at all improbable that her female parent had bred a mule in the first instance, and, as in the well-known cases of mares which have been mated with quaggas and zebras, her subsequent progeny, when mated with a horse, shows some trace of the first union.

Telegony

Tegetmeier & Sutherland were here upholding the concept of telegony and the 'well-known cases' to which they referred were 'Lord Morton's ponies'. Having no knowledge of genetics, nineteenth-century naturalists believed erroneously that a mare's succeeding foals could be influenced by a stallion from an earlier mating. The best known and earliest illustrated example of this supposed influence, known as telegony, was described by Lord Morton in a communication read to The Royal Society on 23 November 1820 (Morton 1821):

Some years ago, I was desirous of trying the experiment of domesticating the Quagga, and endeavoured to procure some individuals of that species. I obtained a male; but being disappointed of a female, I tried to breed from the male Quagga and a young chestnut mare of seven-eighths Arabian blood, and which had never been bred from; the result was the production of a female hybrid, now five years old, and bearing, both in her form and in her colour, very decided indications of her mixed origin. I subsequently parted with the seven-eighths Arabian mare to Sir Gore Ouseley, who has bred from her by a very fine black Arabian horse. I yesterday morning examined the produce, namely, a two years- old filly, and a year-old colt. They have the character of the Arabian breed as decidedly as can be expected, where fifteen-sixteenths of the blood are Arabian; and they are fine specimens of that breed; but both in their colour, and in the hair of their manes, they have a striking resemblance to the Quagga. Their colour is bay, marked more or less like the Quagga in a darker tint. Both are distinguished by the dark line along the ridge of the back, the dark stripes across the fore-hand, and the dark bars across the back part of the legs ... Both their manes are black; that of the filly is short, stiff, and stands upright, and Sir Gore Ouseley's stud groom alleged that it never was otherwise. That of the colt is long, but so stiff as to arch upwards, and to hang clear of the sides of the neck; in which circumstance it resembles that of the hybrid.

In 1821 the Swiss artist Jacques-Laurent Agasse (1767–1849) was paid 60 guineas by the Board of the Museum of the Royal College of

Surgeons and commissioned to paint all the equids involved in 'the unusual circumstances connected with the foals of the Arabian mare in the possession of Sir Gore Ouseley' (Loche & Sanger 1988). The animals were the Quagga, the Arab mare, the hybrid filly, the black Arab stallion (the second sire), the filly (first offspring of the Arab sire), the colt (second offspring) and another foal (the third offspring). Three of these paintings, which are still in the ownership of the Royal College of Surgeons, are reproduced in Figures 3.6, 3.7, 3.8.

The case of Lord Morton's ponies was used by Darwin as support for his theory of natural selection. He argued correctly that the two foals of the Arab sire and dam were striped, not because of telegony, that is the influence of the previous Quagga sire, but because all horses have the an underlying pattern of stripes in their repertoire of development (Darwin 1858, 1868; Gould 1990). It is not at all unusual for the foals of horses to have a black stripe along the back and stripes on the legs. The stripes usually fade away as the horse becomes adult but they can remain and in some breeds, for example the Iceland pony, it is usual for there to be an 'eel' stripe along the back.

Zebra hybrids

In his book on horses, written for the 'Naturalists Library' and first published in 1841, Hamilton Smith quoted Lord Morton's description of what he believed was the effect of the Quagga on the subsequent progeny of the Arab mare and had engravings made from the paintings by Agasse. He was much intrigued by the interbreeding of different species of equids and wrote that:

> Already, in the time of Buffon [1707–88], the idea of producing mules from the striped species of Equidae had occurred. Lord Clive, in experiments to effect this purpose, had found it necessary to deceive a female zebra by painting a male ass with hippotigrine stripes.

The preoccupation of these early nineteenth-century livestock breeders with the inheritance of bloodlines provided Darwin with much of the evidence for his theories. Darwin believed in telegony, or 'the direct action of the male element . . . on the subsequent progeny of the female by a second male.' (Darwin 1868 1). With the elucidation of the laws of inheritance and genetics it is now known that the progeny of one sire cannot be affected by any influence from a previous sire, yet even today many breeders of pedigree animals, especially show dogs, will still try to ensure that their females are not 'contaminated' by a poorly-bred male.

During the nineteenth century numerous experiments were carried out in attempts to domesticate Burchell's zebra (*Equus*

Fig. 3.6 Hybrid foal, which the Arab mare bore when sired by the quagga in Lord Morton's experiment in cross-breeding. Painted by Jacques-Laurent Agasse, 1821. (Photo Royal College of Surgeons of England.)

Fig. 3.7 Filly, the first progeny of the Arab mare when sired by the Arab stallion, painted by Agasse, 1821. (Photo Royal College of Surgeons of England.)

Fig. 3.8 The Arab mare with her third foal by the Arab stallion, painted by Agasse, 1821. (Photo Royal College of Surgeons of England.)

burchelli, the common zebra of Southern Africa) and they were interbred with both horses and donkeys in Europe and in South Africa (Fig. 1.6). It was the view of Tegetmeier & Sutherland (1895) that mules produced from zebras and donkeys would be very successful in South Africa, as they would be immune to horse sickness (an endemic and lethal 'distemper'). But the breeding of zebra hybrids never became an economic proposition, perhaps because the army was too conservative to take on a new kind of pack animal which was striped and might be even more stubborn than the mule. In any case the demand for mules was declining at this time, as they were replaced first by the steam engine and then by the petrol engine.

The bell-mare

For at least 3000 years, until the invention of the steam engine, the mule was the favoured pack animal for long-distance transport, for war, and for agriculture. Mules were uncommon in only a few countries, including the British Isles, but even in England mule-breeding was practised on a small scale from the time of the introduction of the donkey by the Romans. In many parts of the world the breeding and management of huge numbers of mules was an industry which, like any other, had innumerable traditions and conventions. For example it was generally known that the mule team would pull harder and faster if it was lead by a mare. Dent in his book on the history of the donkey (1972) quoted from a description of American mule-trains of the 1900s which were always led by a bell-mare. This is a mare with a bell round her neck who is willingly followed wherever she leads by the mule-train, in the same way as a bell-wether will lead a flock of sheep. As Tegetmeier & Sutherland put it, 'Brought up by the side of the mare (his dam) the mule adores the whole horse tribe, and hates the asinine race generally. He is always nervous, and afraid of strangers'. These authors record an anecdote that is worth repeating:

Some years ago, during the progress of one of the little wars in South Africa, certain 'imperial officers' were sent up the country to buy mules for the service. Arrived at a breeding farm, which happened to belong to an educated English gentleman, certain mules were shown which were running in an inclosure with two old ponies, the latter for company's sake. A bargain was struck for the whole of the mules, and it was suggested by the seller that the officers should take the two old ponies for an old song, as it might facilitate their getting the mules down to headquarters. The seller was rather curtly informed that their 'orders were to buy mules, not ponies.' The absence of any practical knowledge of the subject is as self-evident as is the want of discretionary power accorded to the purchasing officers. It is thought that the mules are still wandering about the veldt somewhere in South Africa.

The 'educated English gentleman' was probably C.L. Sutherland himself who took a great interest in the breeding of mules and who, in the 1880s, presented the skeletons of a female Poitou donkey and a male French mule to the British Museum (Natural History).

Fig. 3.9 Poitou jack donkey. From Tegetmeier & Sutherland (1895).

Poitou mules

The province of Poitou in western France was, for hundreds of years, a centre for the breeding of exceptionally large mules which stood more than 16 hands (163 cm) at the withers (Fig. 3.4). These mules were not pack animals, they were too large for baggage to be lifted on to their backs, but were used for the kind of heavy agricultural work that was done by shire horses in Britain and the Low Countries. The exceptional size of the mules was achieved by crossing the specially bred, very large Poitou jackasses with large mares. Apart from its large size, it is in the feet and limbs that the Poitou mule differs from others, for the legs are short and stout and the hooves unusually wide for a mule. Today in Poitou these huge mules are no longer to be seen working in the fields and only a few Poitou donkeys are kept by enthusiasts of the breed (Fig. 3.9).

The beast of burden

At the beginning of the nineteenth century the enormous potential of the mule for agricultural and draught work in North America, and as a pack animal for colonial armies in many countries of the world meant that there was a great demand for jacks from Europe. The

Fig. 3.10 A twenty-mule team crossing Death Valley, California, in the late nineteenth century. The teams travelled 165 miles from the Harmony Borax works across salt fields, mountains, and desert to the railroad in Mojave. Each waggon weighed three and a half tons and carried nearly ten tons of borax. (Photo Borax Consolidated Limited.)

most popular jacks came from Spain because they were large-boned and produced large and very powerful mules. This account is from the driver of an American mule-train, written just before the new invention of the railway engine relieved these unfortunate animals from what must often have been unendurable burdens (Fig. 3.10):

I have just returned from a trip west with a mule train, of about 400 miles, through a country where bridges are unknown, and the roads are the best place you can find to drive – sometimes mountainous, intersected with steep-banked creeks; at others long steep rises, with draws between 2ft. or 3ft. deep in black mud, and after a rain almost impassable for miles . . . Six mules, the leaders no larger than ponies, will take 6000 to 7000 pounds anywhere, making fifteen to thirty miles a day according to the state of the roads, and I have known a team in summer driven fifty miles with 1000 pounds a head of load, to reach water, and not appear to suffer. They do not require the feed horses do (who invariably lose flesh in the winter time), but will live on maize with very little roughness.

Tegetmeier & Sutherland 1895

PART III
HISTORY OF THE DOMESTIC HORSE, DONKEY AND MULE

4 The earliest domestication of the horse and the ass

The horse

In the legends of classical Greece, the Centaurs were mythical creatures who were half man and half horse (Fig. 8.6 see p. 114). They came from Thessaly and were held by one legend to be the offspring of Centaurus who was the son of Apollo. Another claimed that they were the progeny of Ixion, a king of Thessaly who was seduced by a cloud in the form of Juno.

A more likely explanation is that a group of people from Thessaly, perhaps around 1000 BC, were seen from a distance riding on horses and this unfamiliar scene was perceived as proof that creatures existed which were half human and half horse. To us, the riding of horses, donkeys, mules, camels, cattle, and reindeer is so familiar that it is hard to envisage a world in which it did not occur, and yet it appears that this simplest of all methods of transport had a surprisingly late start. The domestication of goats, sheep, cattle, and pigs began around eight thousand years ago but that of the horse and donkey did not take place until about six thousand years ago.

Fig. 4.1 Ivory carving of a horse's head from St-Michel d'Arudy, Pyrenees. (Photo Jean Vertut.)

Fig. 4.2 Head of a living Przewalski horse in summer coat. (Photo Geoffrey Kinns.)

Towards the end of the last ice age, twelve thousand years ago, the wild horse was an important provider of meat, and probably hides, to the growing human populations of western Europe. It is not impossible that at this time an occasional foal was tamed and ridden, but it was never a widespread practice. In 1906 Piette published an account of a small carved head of a horse that had been found in the cave of St-Michel d'Arudy in the Pyrenees. It had been carved out of mammoth ivory during the Upper Palaeolithic period and it aroused especial interest because Piette claimed that it showed the horse to be wearing a rope halter (Fig. 4.1). In 1978 Bahn raised this question again and postulated that horses were ridden at this time. However, it is just as probable that the artist who made the horse's head some 14 000 years ago was trying to show the mealy muzzle and muscle lines that he saw in the living horse and which can be seen today, for example in an Exmoor pony or a Przewalski horse (Fig. 4.2).

Olsen (1989) in her reappraisal of the remains of the horses found at Solutré in France, examined the long-accepted view that the wild horses were driven to their deaths over a cliff by *mounted* hunters. However, she rejects this hypothesis in favour of the less dramatic

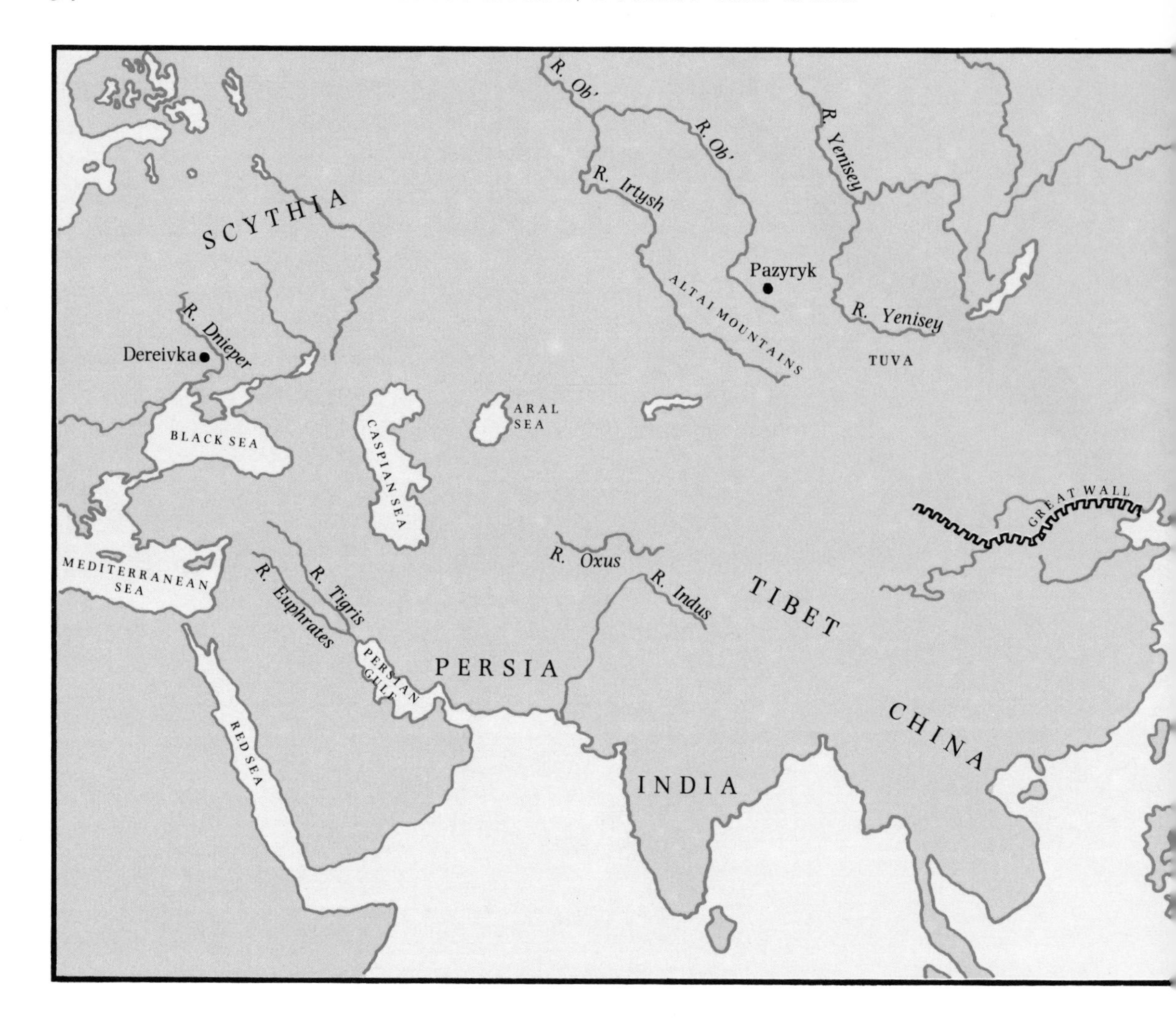

but more rational supposition that the horses were driven by hunters on foot (see p. 24).

Fig. 4.3 Map of Asia with the Neolithic site of Dereivka and the Scythian tombs at Pazyryk.

From 9000 years ago the wild horse became increasingly rare and its remains are seldom found on archaeological sites in Europe, but from around six thousand years ago remains of horse begin to appear in cultural contexts that indicate the beginnings of domestication. There have been sporadic finds throughout Europe but the best known are those from the sites of the specialized Neolithic culture, known as Sredni Stog, in the steppes north of the Black Sea (Fig. 4.3).

During the 1960s and in 1983 extensive excavations were carried out of a late Neolithic settlement that has become well known as the site of Dereivka on the river Dnieper (Telegin 1986, Levine 1990). The bones of at least fifty-two horses were retrieved from this site including seventeen skull parts that were complete enough to be sexed. Of these, two came from mares and fifteen from young stallions. In another Sredni Stog level, near a hearth and a stone structure, a horse skull and limb bones had been buried together with the skeletons of two dogs (Bibikova *in* Telegin 1986, Anthony 1991, Anthony & Brown 1991). The withers height of this horse, as calculated from the lengths of the limb bones, would have been about 144 cm (*c.* 14.5 hands), which means that it was within the range of a present day Przewalski stallion. However, most of the horse remains from Dereivka came from animals that were considerably shorter, with heights of the withers in the range 132–140 cm. The average height of the Przewalski horse today is 138–146 cm for stallions and 134–140 cm for mares (Groves 1974).

It is clear from the remains of these horses at Dereivka and other sites that there was a pattern of specialized exploitation of the horse, about 6000 years ago, in the steppe lands of the Ukraine. Stallions were probably killed for their meat, mares could have been milked, and both sexes may have been used as draught animals, but until recently it was generally believed by archaeologists that there was no evidence for horseriding until the much later time of around 1000 BC. However, Anthony (1991) and Anthony & Brown (1991) argue that it is very unlikely that horses were maintained as a food resource at Dereivka for hundreds if not thousands of years without being ridden, and they believe that riding provides a unified explanation for many of the observed changes in the archaeological record.

That at least some of the horses were driven or ridden with a bridle and bit is evident from the abnormal wear on the premolars of the stallion's skull, and from the finding of six perforated tines of red deer antler at Dereivka that are claimed to be the cheek pieces of bridles (Fig. 4.4; Telegin 1986; Anthony & Brown 1991).

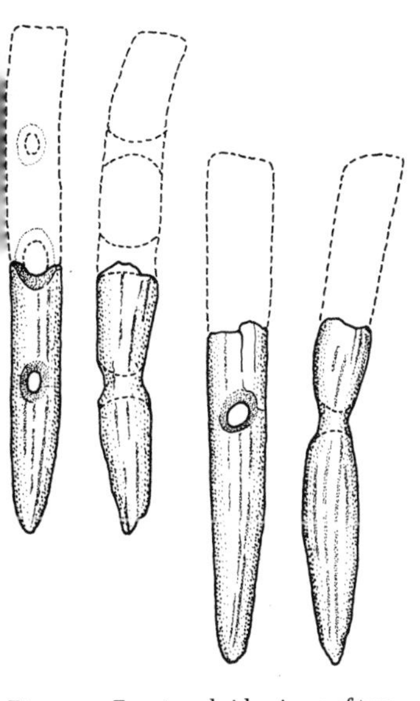

Fig. 4.4 Front and side views of two perforated antler tines from Dereivka. After Telegin (1986).

It has been suggested that domestication of the horse occurred at this time because it enabled the expanding human populations to move away from the river valleys, which were becoming deforested and over-hunted, and into the steppes where the wild horses provided a new resource. At the same time a cold climatic phase (known as the Piora oscillation) encouraged the keeping of horses in this region as they were more able to withstand extreme cold than cattle or sheep (Anthony 1986). Horses will also paw aside the snow to graze, which neither donkeys nor cattle will do.

It must be emphasized, however, that there are very few hard facts to substantiate the 'where and when' of early horse domestication and new evidence is accumulating all the time from the osteological study and dating of excavated material. Levine, in a new appraisal (1990), has suggested that the horse remains from Dereivka may after all be from wild, hunted horses, not from domestic ones, and Uerpmann (pers. comm.) believes that horses were locally domesticated from relic populations in many parts of Europe during the Neolithic period.

Once the horse had been domesticated its use for draught spread very quickly into western Asia and eastern Europe. In both these regions the evidence for wild horse in the early Neolithic period has been controversial. It was previously assumed that there was no wild horse in western Asia after the end of the last ice age, but one record from the fourth to early third millennium site of Norsun Tepe on the Altinova Plain in Anatolia was interpreted by Boessneck & Driesch (1976) as coming from the wild horse. However, Bökönyi (1991) who has described further finds from Chalcolithic sites on the Altinova Plain, believes them to be from domestic horses. From further east and south, within the last few years, a few horse bones have been identified from the Chalcolithic sites of Shiqmim and Grar in the northern Negev region of Israel, dating to the fourth millennium BC (Grigson 1992). As these are very unlikely, on biogeographical grounds, to be from wild horses, the indications are that all the horse remains of this period from Anatolia are also from very early domestic animals.

Further west, in eastern Europe, Bökönyi (1974*b*) claimed that there were no remains of horses in Hungary until domestic equids were introduced in the Bronze Age (during the second millennium BC), but later Vörös (1981) identified wild horse from a number of early Neolithic sites in the Carpathian Basin.

By 2000 BC, while the wild horse continued to be pushed into its eastern refuges, the domestic horse had begun to spread rapidly around the whole of the Old World, with finds becoming common on late Neolithic sites throughout western Asia and right across Europe to the British Isles. Most of this newly domesticated stock was probably derived from the core area north of the Caspian Sea, but it is not impossible that some local domestication occurred from the dwindling herds of wild horses in several parts of Europe. In Asia, from south of the Caspian Sea at Tal-e Malyan in southern Iran (Fig. 4.5) remains of horse have been found dating to 2000–1800 BC (Zeder 1986).

In this first phase of domestication, these early horses do not appear to have become much smaller than their wild progenitors, which is what happened to all the other common domestic animals from dogs to cattle when they were first bred in captivity (Clutton-Brock 1987). The remains of the earliest domestic horses from Dereivka are claimed, however, to show more variation in size than would be expected in a wild population.

It is very difficult to investigate the osteological changes that took place in the skeletons of the earliest domestic horses because there are so few fossil records of the wild horse, dating to the fifth and

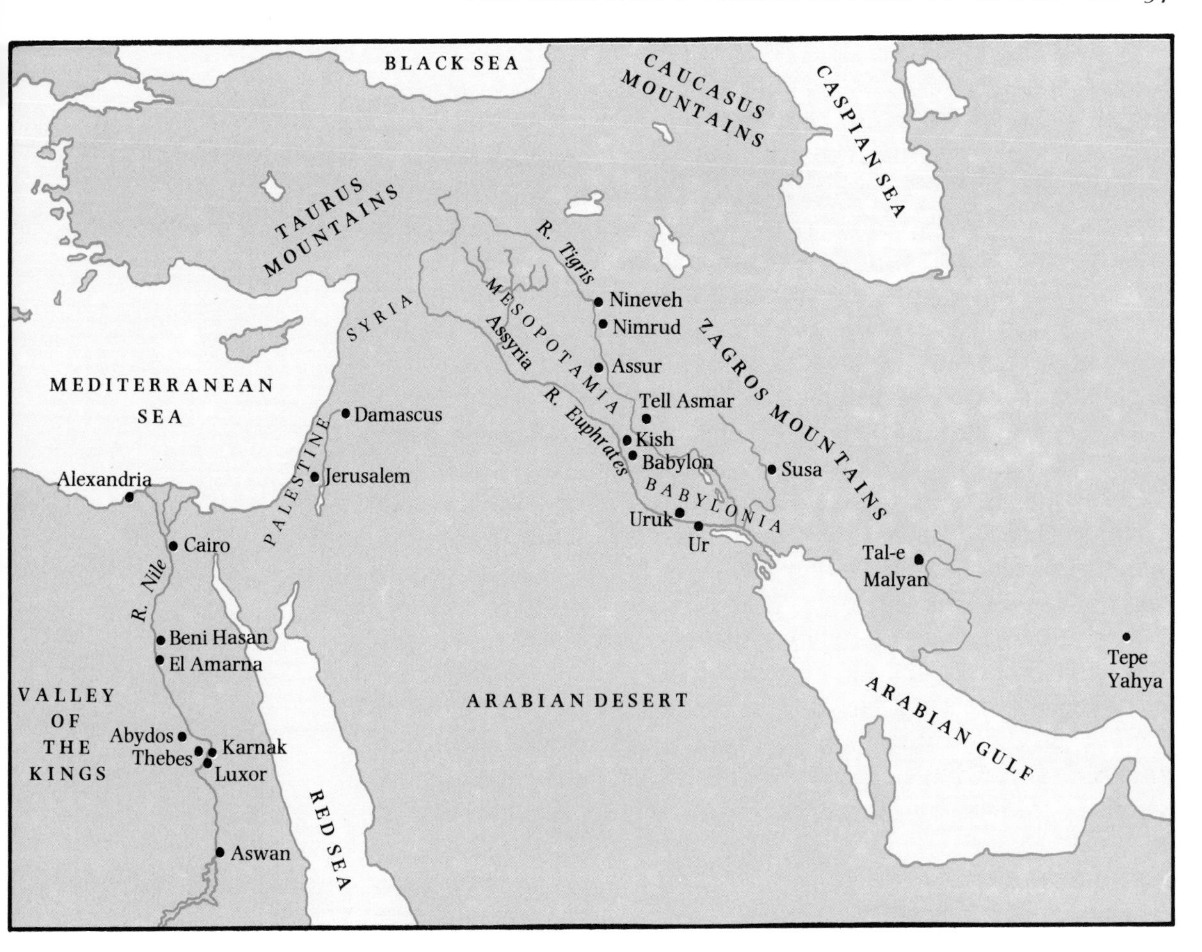

Fig. 4.5 Map of Egypt and western Asia.

fourth millennia BC, with which they can be compared. With cattle, for example, very large numbers of bones and teeth of *Bos primigenius* (the wild ancestor of domestic cattle) have been retrieved from Neolithic sites, so the transformation from wild to domestic can be traced in the osteological record (Clutton-Brock 1987). But the wild form of the horse had become extinct over nearly all its late Pleistocene range in Europe and western Asia several thousand years before the first domestic horses were introduced from the eastern steppes. In Britain, for example, bones of the wild horse are found on most Late Glacial sites, as at Gough's Cave at Cheddar in Somerset (Burleigh 1986), but after about 7000 BC there are no remains of wild horses that have been authenticated by direct dating.

In western Europe it is proving difficult to determine whether the few isolated finds of horses from the earliest Neolithic are from the relics of wild populations or early examples of domestic horses. To make the problem more complicated, when radiocarbon dates are obtained on these remains they often turn out to be either earlier than the Neolithic, as in the case of a horse bone from Kaster, an early Neolithic site in Germany, which gave a date of 10 380 ± 140 BP* [OxA-1392†], or later as in the case of a bone from another Neolithic site in Germany, Lonsberg, which gave the relatively modern date of 1030 ± 90 BP [OxA-1393] (Hedges *et al.* 1989).

In Denmark there is incontrovertible evidence for the remains of horses in the Middle Neolithic period. Horse bones were discovered in three pits at Lindskov in 1899, and nearly 100 years later a radiocarbon date was obtained from one of the bones of 2570 ± 65 BC [k–2652] (Davidsen 1978). The archaeological site that yielded these horse remains belongs to the Funnel-Beaker culture (or Trichterbecher, which is usually abbreviated to TRB), and although it is not impossible that these people had domestic horses Davidsen (1978) was more inclined to believe they were wild.

At the site of Moncín in Spain, direct dating of remains of horses which had been assumed to be from wild animals dating to the early Neolithic period, found that they belong in fact to the Late Bronze Age (3080 ± 120 BP [BM-2193R]). This means that they are most probably from domestic animals. In other parts of Europe small numbers of horse remains have been identified from Neolithic sites, for example in the Netherlands (Clason 1986), Switzerland (Clutton-Brock 1990), and Hungary (Vörös 1981). But the absolute age of these finds should be determined by radiocarbon dating of the actual bones and teeth before assertions are made about their status as wild or domestic horses.

Perhaps the least controversial early record for the domestic horse in western Europe comes from Ireland where horse remains have been retrieved from the Late Neolithic (Beaker) levels of the site of New Grange in County Meath (Wijngaarden-Bakker 1974). As there are no bones of horse in the fossil record of Ireland it can be assumed that these remains must be from domestic horses that were brought into Ireland around 2000 BC which is the date when horse begins to appear with some frequency on archaeological sites all over Europe and western Asia. From the mainland of Britain a number of what appear to be ritually buried skulls of horses have been retrieved from around this date. One of these skulls, from the Neolithic flint mine site of Grime's Graves in Norfolk has been radiocarbon dated to 3740 ± 210 BP [BM-1546]. It was a particularly interesting find because the skull was from a very aged mare who had lost almost all her teeth and would not have been able to obtain food by grazing. She must therefore have been a highly valued animal that was specially fed until she died (Clutton-Brock & Burleigh 1991).

Throughout Europe and western Asia, by 2000 BC, the horse was slowly replacing the ox as a much faster means of pulling carts. In western Europe the remains of horses are associated with the Beaker culture of the Late Neolithic/Early Bronze Age people who were

* BP = Before Present
† This is the reference number for the radiocarbon date, with the prefix indicating the laboratory where the date was obtained.

setting up new exchange networks in connection with the developing technology of metallurgy (Renfrew 1987). It was not, however, until more than 1000 years later that incontrovertible evidence for horse-riding becomes commonplace. One famous example is the depiction of warriors on horseback on the scabbard from a Hallstatt grave which Powell (1971) ascribed to the seventh century BC (Fig. 4.6).

Fig. 4.6 A mounted warrior engraved on the scabbard from grave 994, Hallstatt, Germany. From Powell (1971).

The early writers on the history of the horse assumed that because, since ancient times, there have been different forms of domestic horses, with small stocky ponies in the north, heavy horses in middle Europe and the Arab in Egypt and western Asia, these must have come from different ancestral races. Ewart (1907), for example, postulated that there was a *Steppe* variety, allied to Przewalski's horse which was long-headed, a *Forest* variety which gave rise to the ponies with short broad heads, and a *Plateau* variety in which he placed the horses with long narrow heads. This category included the Libyan (Arab) as well as the Celtic ponies. Ridgeway, whose influential book *Origin and Influence of the Thoroughbred Horse* was first published in 1905, followed Ewart, for whom he had a great admiration, in believing that the progenitor of the Arab was an endemic wild horse which inhabited Libya in the prehistoric period.

Undoubtedly there were wild equids in North Africa during the Pliocene and early Pleistocene. Apart from the ass, there were ancestral zebras, for example *Equus burchelli mauritanicus*, and the three-toed equid, *Hipparion libycum* which survived into the Upper Pleistocene. But there is no evidence for the presence of the true horse, *Equus ferus*, in the fossil record of North Africa (Churcher & Richardson 1978).

At the same time as Ewart and Ridgeway were putting forward their view that the Arab horse had an endemic evolution in North Africa, Lydekker, the well-known palaeontologist, contended that

Fig. 4.7
Shetland ponies.
(Photo
Sally Anne Thompso

Fig. 4.8
Arab horse.
(Photo
Sally Anne Thompso

the Arab was descended from the fossil horse of India, *Equus sivalensis*. Ridgeway thought nothing of this and wrote (1905):

> Nor is it only in colour and other external respects that the Libyan differs from the Asiatic horse. As the cry of the quagga, from which that animal derived its name, was distinct from that of the zebra, so the voice of the Libyan horse differs from that of his vulgar Asiatic brother.

Such is the manner of scientific discourse, and indeed there is justification in studying the vocalizations of animals as a clue to their taxonomic relationships, but the truth can only be learned from a combination of investigations into fossil histories, osteology, molecular biology, and studies of natural behavioural patterns. In the case of the horse the accumulated evidence indicates that all domestic horses of the past and present are descended from the single ancestral species, *Equus ferus*, with the focus of its domestication being on the Scythian steppes, north of the Caspian Sea (Ukraine). The division of the domestic horses, into the so-called 'hot-blooded' and the 'cold-blooded', is a reflection of the variation in morphology that can be seen in many wild and domestic species of mammals which have a very widespread distribution. In northern latitudes and cold climates mammals tend to be large (a generalization that is often called Bergmann's 'rule') and have heavy bodies with short legs, while the extremities, such as the ears, are short and compact (Allen's 'rule'). In their range over lower latitudes with hotter climates the same species of mammals tend to have longer and finer limbs relative to the size of the body and longer extremities (Figs. 4.7 & 4.8). The hot-climate mammals will also have a shorter and sleeker coat. It can be seen that, despite artificial selection for characters other than those required for adaptation to special environments, many breeds of domestic livestock fall into this pattern. Compare for example the long-legged, lop-eared Persian breed of sheep with the Lincoln longwool of Britain, both of which are descended from the same wild progenitor, the Asiatic mouflon, *Ovis orientalis*.

When looked at in terms of biological adaptation, it is entirely logical that the horses of the deserts of Arabia and North Africa should have evolved into elegant fine-limbed animals while the northern ponies retained the heavy head, shaggy coat, and relatively short limbs of the tarpan. The terms, 'hot-blooded' for the Arab, 'cold-blooded' for the heavy draught horses, and 'warm-blooded' for the three-quarter bred horse may be useful terms for the horse-breeder but they hold no scientific credence. All domestic horses have the same body-temperature and all the different breeds will mate and produce fertile offspring.

The early writers on the history of domestic horses and ponies were not only interested in the size and conformation of the various

Fig. 4.9 Dun Devonshire pony with shoulder, spine, and leg stripes. From Darwin (1858).

breeds, they were also much intrigued by the variation in coat colour, and by the sporadic occurrence of a longitudinal dark band along the ridge of the back and stripes on the shoulders and front legs (Chapter 3). Darwin collected examples of shoulder and leg stripes in horses of very different breeds in various countries from Britain to eastern China, and he asserted that the stripes occurred most frequently in horses of dun and mouse-dun colour (Darwin 1858). This is perhaps not unexpected as the mouse-dun colour does appear to be the wild-type, as can be seen from the Przewalski horse of today and as described for the wild tarpan by Gmelin (1770). The appearance of stripes on the legs and other parts of the body of horses and mules may occur as a reversion to an ancestral equid characteristic which, like the mealy muzzle of the wild horse, has been lost in most breeds of domestic horses (Fig. 4.9).

Darwin believed that all domestic horses were descended from 'a single, dun-coloured, more or less striped, primitive stock, to which our horses still occasionally revert.' (Darwin 1868 1); a conclusion against which there can be little argument today.

The ass

Unlike the horse which can survive in almost all parts of the habitable world, the donkey (*Equus asinus*), which has a desert

Fig. 4.10 Somali wild asses *(Equus africanus somaliensis)* that were taken from the Danakil region of Ethiopia to the Hai Bar Yotvata Reserve in Israel. (Photo H. Mendelssohn.)

origin, is more particular in its needs. Even so, the donkey can flourish as successfully on the damp, peaty farms of Ireland as it can in the countries of the Mediterranean. This is because different races have evolved over the last 2000 years by natural selection in response to local conditions of climate and nutrition.

Only one species of equid has been the progenitor of the domestic donkey, this being the wild ass, *Equus africanus*, (Fig. 4.10) which was once widespread as a number of separate subspecies over the whole of Saharan Africa and most probably also in Arabia (see Chapter 2 & Appendix). The wild ass is an elegant, fine-limbed equid that can gallop at great speed, and therefore the evolution of the different races of domestic donkey present an opposite progression from that of the horse. The natural habitat of the wild horse was cold, steppe-grasslands and consequently it was a stocky animal with short ears and a heavy head. However, when bred in the hot arid regions of North Africa and Arabia the domestic horse soon

Fig. 4.11 Domestic donkeys in Ira
(Photo author.)

developed into the fine-limbed Arab breed. The donkey, on the other hand, began as a fine-limbed 'Arabian' type which became more stocky as it moved north, so that in the short-limbed, heavy-headed, shaggy, donkey which is so distinctive of Ireland today, only the large ears and small hooves remain to witness its desert origin (Figs. 4.11 & 4.12).

The fossil record of the wild ass (*Equus africanus*) in the prehistoric period is sparse and poorly documented with no bones that have been directly dated. In Algeria, Thomas (1884) described the remains of wild ass that are assumed to be early Holocene and which he named *Equus asinus atlanticus*. From Upper Egypt there are a few records of wild ass of Late Pleistocene and early Neolithic age (Churcher & Richardson 1978), and there are also a few tenuous records from the Arabian peninsula and Syria (Uerpmann 1987).

From the archaeological record as well as from ancient art and from textual evidence it is clear that the wild ass was first bred in captivity in Egypt and in western Asia and that it was in general use, presumably as a beast of burden, by 2500 BC. In Egypt the skeletons of three donkeys were found by Sir Flinders Petrie in his excavations carried out at the beginning of this century in the tomb of Tarkhan (Petrie 1914). A radiocarbon date of 4390 ± 130 BP [OxA-566] has recently been obtained on bone from one of the skulls. From northern Sudan, Peters (1985–6) has identified a foot bone of a domestic donkey from the Neolithic site of Jebel Shaqadud.

In western Asia domestic donkey has been identified from a number of early sites as far east as Tal-e Malyan in southern Iran (Fig. 4.5) where Zeder (1986) has postulated that donkeys may have been in use as early as 2800 BC. From this time onwards the numbers of domestic donkeys represented on archaeological sites increases greatly, as does the evidence from Sumerian records written on clay tablets. By 1000 BC it is evident that the domestic donkey was the common means of transport throughout Egypt and western Asia, as was the horse in the rest of Asia and Europe.

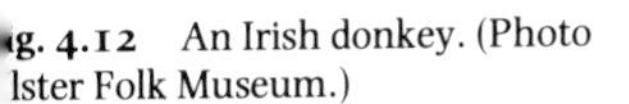
ig. 4.12 An Irish donkey. (Photo lster Folk Museum.)

The pictorial evidence from the third millennium for the riding of both horses and donkeys usually shows the rider sitting far back in the 'donkey seat' (Moorey 1970; Fig. 4.13). Because of its low withers and low carriage of the head and neck the only way to ride a donkey for any length of time is to sit well back on the loins, but even then the rider will be jolted by the shocks of the moving legs if going at any speed. For this reason the donkey is a quite unsuitable animal for riding at speed and this would have precluded its use as a mount in hunting or battle.

Fig. 4.13 The 'donkey seat'. Drawing of a terracotta plaque. From Littauer & Crouwel (1979) after Moorey (1970).

5 The first wheeled transport, horse-riding, the stirrup, and nose-slitting

The need for transport and traction

Humans, like very many species of animals, live in a home which they consider to be their own territory and is the centre of their home range. In a foraging or hunting community this home range will probably have boundaries that are within a comfortable walking distance of the home. When the home is moved, whether often or seldom, it involves travel, either within the known home range or outside it.

Since their first evolution from early hominids, humans have needed to move around in order to obtain food, but to leave the home range and venture into unknown lands, though a perilous enterprise, has always been irresistible to some individuals. In the modern world it is hard to understand how a human group could survive without belongings and without travelling outside the home range, yet this was the natural way of life for most of the span of human existence.

Once this is understood it does not seem so strange that people took so long to learn to exploit animals as beasts of burden. Until the discovery of metals the majority of people had few permanent belongings, and usually there was no need to travel far. However, with the development of larger settled communities in the late Neolithic period, everything changed; dwellings became more permanent and contained furniture. The economy changed from self-sufficiency to trade in food and goods that had to be brought into the towns. A cart or sled drawn by domestic cattle, which for 2000 years had already been associated with human societies, was not long in development. Even so, to construct the simplest sled and to harness it to a pair of oxen without the use of metals is no mean feat. It would require a large quantity of wood and leather and would need considerable skill at carpentry, as well as in management of the animals.

The use of animals for pulling a simple plough began at much the same time as their use for transport of goods, but here again the plough ox came before the plough horse (Piggott 1983). The Romans used donkeys for ploughing and grinding corn but in northern Europe horses were not generally used for ploughing until the early Medieval period.

The first carts and waggons

The earliest evidence for the use of carts and for the simple traction plough comes from the pictographs from Uruk in southern Mesopotamia (Iraq) (Fig. 5.1, see Piggott 1968, 1983; Littauer & Crouwel 1979). Both sleds and wheeled carts are represented in these small engravings on clay tablets which have been dated to between 3200–3100 BC. The carts were probably drawn by oxen that were harnessed by a bar across their horns as on the model shown in

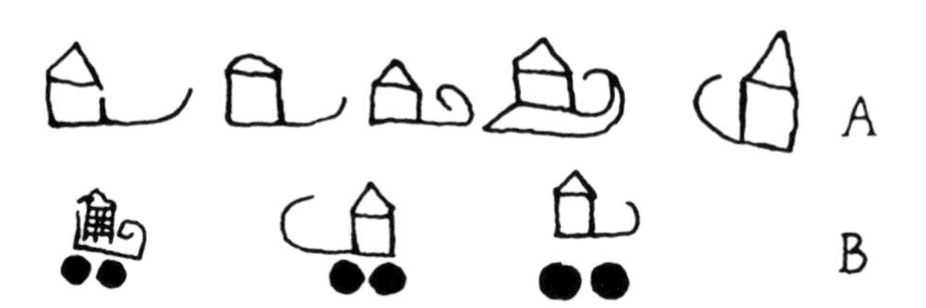

Fig. 5.1 Pictographs of wheeled carts from Uruk, Iraq. From Piggott (1983)

Fig. 5.2 Copper model of an ox-waggon from Maraş, eastern Turkey. Early second millennium BC.

Figure 5.2. The restored ceremonial sledge of Shubad (Puabi) that was excavated from the royal tombs of the Sumerian city of Ur (Fig. 4.5) together with two cattle skeletons shows very clearly what these first vehicles were like (Fig. 5.3). It is dated to the Early Dynastic period (*c.* 2500 BC).

By this period four-wheeled waggons drawn by two or four asses were also in use, as so clearly depicted in the 'battle scene' on the Standard of Ur (Fig. 5.4). The waggons were made entirely of wood, including the wheel axle. They were drawn by two draught animals (polers) harnessed to a central pole, and there was, also, often an

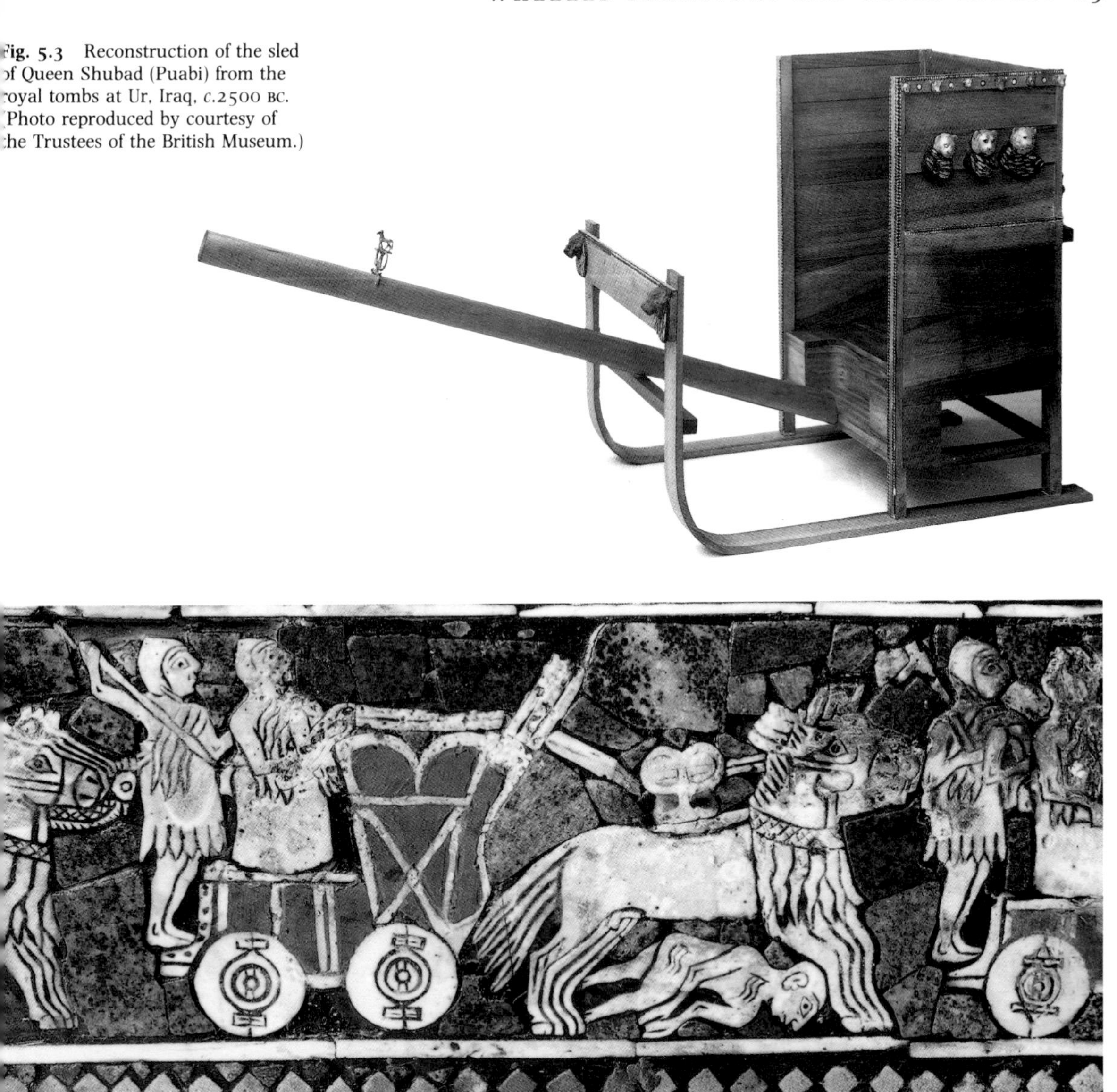

Fig. 5.3 Reconstruction of the sled of Queen Shubad (Puabi) from the royal tombs at Ur, Iraq, *c.*2500 BC. (Photo reproduced by courtesy of the Trustees of the British Museum.)

Fig. 5.4 Battle scene on the wooden box known as the Standard of Ur. (Photo reproduced by courtesy of the Trustees of the British Museum.)

outside pair of asses (outriggers) that were attached by their collars to the collars of the polers. There is no evidence that the earliest carts had the swivelling front axle that enables a four-wheel cart to turn corners. This invention is first reported from the Hallstatt (Iron Age) period (Piggott 1983). The lack of a swivelling axle probably explains why so many of the vehicles in the ancient world only had two wheels. A four-wheeled chariot would be extremely unstable and would tip over as soon as the horses turned a corner at speed. So after this initial trial by the Sumerians the two-wheel chariot and baggage cart became the normal means of transport in peace and war.

Furthermore, the animals that drew these vehicles had always to be in pairs because the only known method of harnessing was with a central pole. The idea of putting a single draught animal between shafts does not seem to have occurred to anyone until around the second century BC, when shafts are shown in Chinese documents of the Western Han period (Spruytte 1983*a*).

Wheel-making

The earliest wheels were made of solid wood, but as described by Piggott (1968) a circular cross-section of a tree trunk could not be used because it would split. It was necessary to construct the wheel out of longitudinal sections of the trunk. The usual method was to use three planks of wood joined together with struts, the outer two

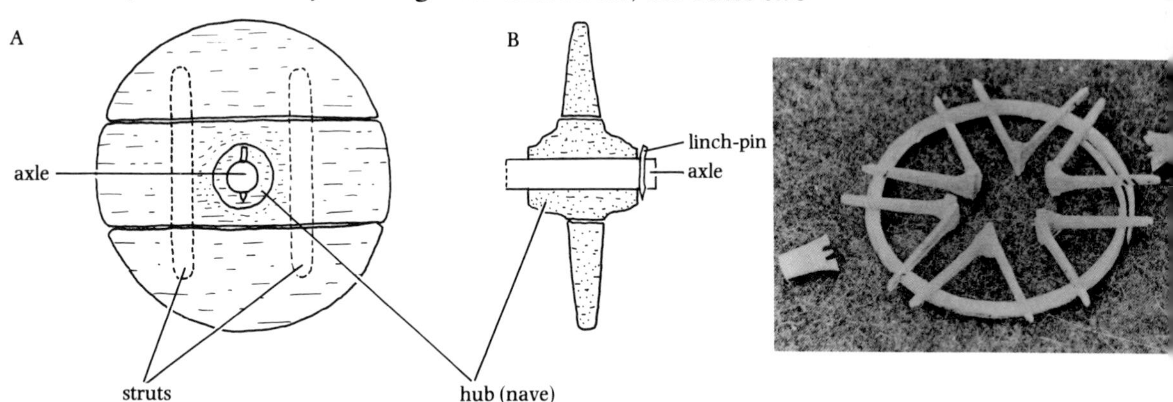

Fig. 5.5 Construction of a solid wooden wheel. A: front and B: side view. After Piggott (1968).

Fig. 5.6 Reconstruction of the spoked wheel from the chariot in Tutankhamun's tomb. From Spruytte (1983*a*).

being half as wide as the central plank (Fig. 5.5). The hub, or nave, through which the axle passed at the centre of the wheel was either a simple hole or it was left as a thicker part when the central plank was shaved flat, or it was a separate tube of wood which was inserted and fixed into the central hole. The hub was held on to the axle with a peg or linchpin on the outside of the wheel.

Spoked wheels are first found from eastern Anatolia (Turkey) around 1900 BC and were common in ancient Egypt. However, these wheels were constructed in quite a different manner from present-day spoked wheels. As described by Spruytte (1983*a*) who made a reconstruction of the chariot found in Tutankhamun's tomb, the Egyptian wheel was made by putting together six chevron-shaped pieces of wood (Fig. 5.6) which were attached to the hub by glue and by 'green' hide that was shrunk on to it. These spokes were then fixed to a rim of bent wood that had a 'tyre' of hide around it. This construction may have been quite efficient in the arid climate of Egypt, but it would not have held together in the damp air of Europe where the cross-bar wheel shown in Figure 5.7 (Littauer & Crouwel 1977) has had a continuity of use since the prehistoric period,

ig 5.7 Cross-bar wheel in England, nineteenth century. From Littauer & Crouwel (1977).

presumably because it was lighter than the disc wheel but more robust than the spoked (Cotterell & Kamminga 1990).

Harness

The ancient chariot or cart consisted of four parts; a box or basket which held the passenger or goods, an axle, two wheels, and the pole for harnessing. It is likely that all systems of harness devolve from the earliest use of cattle for traction where the animals were attached to the pole by a yoke that was either tied to their horns or placed over their withers. In the earliest equid harness depicted from Mesopotamia there appears to have been a similar yoke placed over the withers (Littauer & Crouwel 1979). This would have been held in place by a strap around the neck which in an ox would act to transmit the pulling power efficiently, but on an ass or a horse, which has a longer and thinner neck than an ox, it would constrict the throat.

In 1931 Lefebvre des Noëttes published a book which profoundly influenced many later writers on ancient methods of traction and harness. He maintained that this inefficient method of harnessing with a neck-collar was used throughout the ancient world until the invention of the modern horse-collar in early Medieval times, which he claimed transformed the economic development of Europe. Noëttes, however, failed to understand fully the ancient systems of harness which have been studied in greater detail in recent years, notably by Littauer.

There are only two parts of the body of an equid from which muscular power can be transmitted into traction, the shoulders and the chest, and in both ancient and modern haulage there are two corresponding types of harness to make use of this power (Littauer & Crouwel 1979, Spruytte 1983*a*). In the neck-yoke system the effort of pulling is made by the shoulders while in the dorsal-yoke system the pull comes from the chest. Spruytte maintains that both systems were in use in the ancient world as at the present day (Fig. 5.8A–E).

Although the neck strap may not have helped the equid to make as full use of its pulling-power as the modern shoulder collar there were many other interdependent factors in the manufacture of the haulage cart that counted against its efficiency. These included the fixed wooden axle and the difficulty of constructing light but robust wheels that had to travel over roads which could be deep in mud or sand.

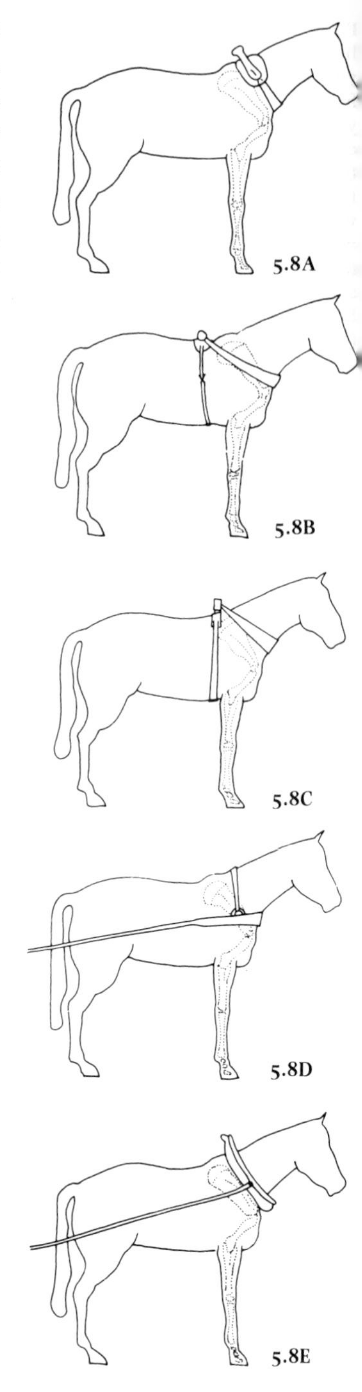

Control of the equids

As with the yoke, the earliest system depicted in Mesopotamia for controlling harnessed equids was adapted from that used for draught cattle; it is shown as single reins attached to a noseband (Fig. 5.9). In the earliest depictions four asses are often shown in harness, side by side, in which case only the two central equids were yoked and the two outsiders were controlled by the reins or traces alone. This tradition continued into classical times when Greek and Roman chariots usually had four or more horses; one Roman charioteer being recorded as winning nine races with a ten-horse team. The reasons for having more than two asses or horses was that, being herd animals, the outsiders encouraged the yoked equids to move faster despite the discomfort and weight of the harness, and a war chariot looked more powerful and was better protected with a large number of horses.

From 1600 BC the light chariot with spoked wheels, but still without a pivoting axle, had become fully developed and is shown in a great many engravings, wall paintings, models, and relics of actual chariots in Egypt and western Asia. It was drawn by two or more horses which were controlled by reins and a bridle with a bit, which was usually made of bronze. Bits were always based on the pattern of the snaffle (a straight or jointed bar), but they became very complicated and later had very cruelly spiked metal cheekpieces

Fig. 5.9 Harnessing by means of reins attached to a nose band. Drawing from the Standard of Ur, from Littauer & Crouwel (1979).

Fig. 5.8 Ancient and modern harnessing systems. A: ancient neck-yoke in which the effort of pulling is from the shoulders; B: ancient dorsal-yoke in which the effort of pulling comes from the chest; C: ancient traction harness as envisaged by Lefebvre des Noëttes (1931), in which the effort of pulling was assumed to come inefficiently from the lower part of the neck; D: modern shoulder-collar harness; E: modern breast-collar harness. From Spruytte (1983*a*).

(Littauer 1969*a*). These were probably necessary in order to control the direction of the chariot horses.

From its beginnings in Mesopotamia the chariot spread all over the ancient world, and for two thousand years it was essential equipage for every king, warrior, hunter, and sportsman of high social status.

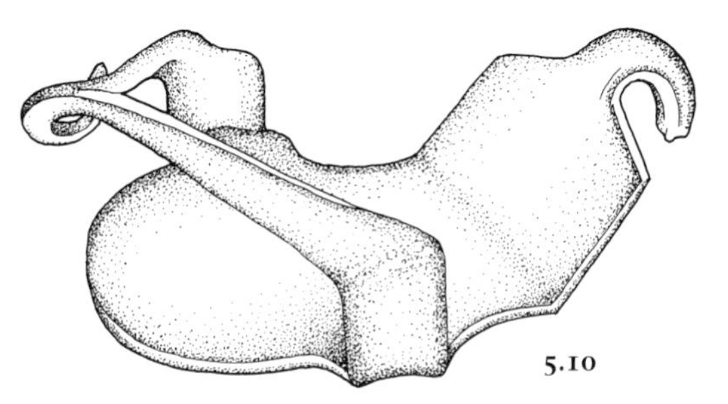

Horseshoes

The equid hoof, like the human fingernail, is made of keratin and grows continuously. On hard, dry ground it will remain healthy and in good condition and will be worn down evenly. But on soft, wet ground it will grow unevenly and will break so that the animal becomes lame and unfit for traction or riding. This is a particularly serious problem in the cold wet climate of northern Europe.

Columella (before AD 70) recommended in his classic work on agriculture (Forster & Heffner 1968) that 'slippers' of broom should be put on to the hooves of oxen that were inflamed, and it is probable that the hooves of horses and mules were similarly treated. The earliest horseshoe was probably the Roman hipposandal (Fig. 5.10), which was made out of a metal sole tied to the hoof with leather straps. White (1962), in support of numerous other authorities, claimed that there was no sound evidence for nailed horsehoes before the end of the ninth century AD when they have been recorded from the Yenisei region of Siberia as well as from many European countries. By the early Medieval period it is probably safe to say that most working horses in northern Europe were shod.

Horse-riding and the stirrup

Around one thousand BC is usually taken as the period when horse-riding became ubiquitous. Before this date there is some evidence for riding in the form of wall paintings and models from Egypt and western Asia, but in these centres of civilization, where the draught ass was an economically more important equid than the horse, riding was probably looked on as an activity of barbarians. The first horses to be ridden had a bridle with a bit but no saddle and no stirrups, and this was how most horse-riding remained, in peace and in war, until the early Middle Ages.

Fig. 5.10 A Roman hipposandal, assumed to be from London, 1st-4th century AD. Length 12.5 cm, width 9 cm, height 6.5 cm. In the British Museum.

Spruytte (1983*a*) proved with his reconstruction of Tutankhamun's chariot that he could keep his balance while the horses were galloping and he did not even have to hold on to the front rail. This indicates that it was probably not too difficult for an archer, as shown in numerous ancient Egyptian paintings, to shoot arrows from a bow while standing in a chariot. However to shoot arrows while riding and steering a galloping horse, without a girthed saddle or stirrups, would appear to be a far more difficult feat. Here again Spruytte (1983*b*) has shown how it was done, this time by the ancient Assyrians in the seventh century BC. By copying the bridle

Fig. 5.11 The Assyrian method of shooting an arrow from a galloping horse, as demonstrated by J. Spruytte. (Photo J. Spruytte.)

shown on stone reliefs as in Figure 6.2, Spruytte was able to release the reins while shooting an arrow from a galloping horse. The reins, instead of being held in the hands, were attached to a neck collar and were held down by a weighted 'pom-pom'. When it was necessary to direct the horse, the collar was pulled to the right or left and then released again with, presumably, pressure also being applied by the rider's legs (Fig. 5.11).

Later, during the middle of the third century BC, the prowess of the mounted bowmen of Central Asia became legendary and still remains with us as, for example in the term 'Parthian shot'. Parthia was an ancient empire that stretched from the Caspian Sea eastwards to the Indus and its people were deemed to be the most expert horsemen of the ancient world. They won many victories by

ig. 5.12 The Parthian shot. Bronze gurine on the edge of an Etruscan ase. (Photo reproduced by courtesy f the Trustees of the British 1useum.)

shooting their arrows backwards as they galloped away from a battle in pretended retreat (Fig. 5.12). In fact, because of the direction of the propulsive forces, it may be easier to stay on a galloping horse's back, without stirrups, when shooting an arrow backwards rather than forwards.

The saddle was preceded by a cloth which was placed on the horse's back, but it had no girth as can be seen from many classical representations of horse-riding. Without a girth, stirrups could not have been kept in place, and it is therefore reasonable to assume that stirrups would not become an aid to riding until the saddle with a solid framework (treed) and girth came into use. Although the cavalry soldiers of the late Roman Empire did not have stirrups they did ride on a treed saddle that had horns or pommels against which the rider could brace himself (Hyland 1990).

There is no word in ancient Greek or Latin for the stirrup and in Europe the earliest finds of the metal stirrup are from seventh century AD graves in Hungary (Littauer 1981). They appear to have come west from north-east China where stirrups of the present-day type are recorded for the first time from tombs of the fourth century AD. However, much earlier than this, experiments were clearly made with soft straps, as shown on a second century BC stone plaque from India, but these may have been more an aid to mounting than to steady the rider. The etymology of the word 'stirrup' supports this view, for as pointed out by Ridgeway (1905) it can be derived from the early English *stige-rap* where *stigan* meant 'to mount' and *rap* meant 'rope'.

It may be that the Scythians, the nomadic horsemen of the steppes who were described by Herodotus (born 484 BC), experimented with stirrups, because one of the terminals of a gold torque from a fourth century BC kurgan at Kul Oba in the Crimea, clearly shows a rider with his foot in a metal hook attached to a linked metal chain (Fig. 5.13 see p. 82; Littauer 1981; Clutton-Brock 1987). However there is no mention of any form of stirrup (or incidentally of spurs) in the famous treatises on hunting and horsemanship by the Greek writer, Xenophon (died 359 BC).

The most notable modern account of the history of the stirrup is that of White (1962) who reasoned that stirrups first appeared in western Europe in the early eighth century AD. White, in a detailed and closely argued case, proposed that by enabling cavalry to take part in mounted shock combat the art of warfare was transformed and that this was a major factor in the development of feudalism. He wrote (1962):

> The Anglo-Saxons used the stirrup, but did not comprehend it; and for this they paid a fearful price. While semi-feudal relationships and institutions had long been scattered thickly over the civilized world, it was the Franks* alone – presumably led by Charles Martel's genius – who fully grasped the possibilities inherent in the stirrup and created in terms of it a new type of warfare supported by a novel structure of society which we call feudalism.

This was a neat theory that received wide acceptance when it was first published, but has since been much disputed because there is plenty of evidence both from the ancient world and, at a much later date, from the native American horsemen (described in Chapter 10) that it was perfectly feasible to shoot arrows from horseback without the aid of stirrups. On the other hand it can be argued that, whereas men 'born and bred' in the saddle could manage a bow and arrow very competently, the large majority of European soldiers, who perhaps only learnt to ride when adult, would not easily gain this proficiency. There would be a very great difference in riding skills between nomadic horsemen such as the native Americans and the

* The name 'Frank' as used by Medieval historians has two meanings. First, as used here, it means a person of the Germanic nations that conquered Gaul in the sixth century AD and from whom the name 'France' derives. Second, it was the name given in western Asia, during the Crusades, to anyone of west European origin.

peasant farmers of early Medieval Europe who, for one reason or another, enlisted as mounted soldiers.

Fig. 5.14 A horse with slit nostrils on a stone relief from Amarna, c. 1350 BC. (Photo Agyptisches Museum, Staatliche Museen Preussischer Kulturbesitz, Berlin.)

Slitting the nostrils of equids

For more than 3000 years the care of asses and horses has involved innumerable beliefs and practices that have been based on myth and tradition. They range from the idea, described by the Roman writer Columella (Forster & Heffner 1968), that mares could be impregnated by the wind, to blistering for the cure of tumours as described by the Victorian veterinary surgeon, Youatt (1846). But few of these practices have been more widespread or had a greater continuity over time than the slitting of the nostrils, which appears always to have been carried out for the purpose of making the equid 'breathe more freely'. The practice has been rather ignored in the literature on equids, but its history was reviewed by Littauer (1969*b*) who considers that it began in an effort to compensate for impaired breathing caused by pressure from the dropped nose-band that preceded the bit. In the words of Youatt (1846):

> The inhabitants of some countries were accustomed to slit the nostrils of their horses, that they might be less distressed in the severe and long-continued exertion of their speed. The Icelanders do so at the present day. There is no necessity for this, for nature has made ample provision for all the ordinary and even extraordinary exertion we can require from the horse.

The earliest pictorial evidence for slit nostrils comes from ancient Egypt where it is seen on a number of stone reliefs as in Figure 5.14

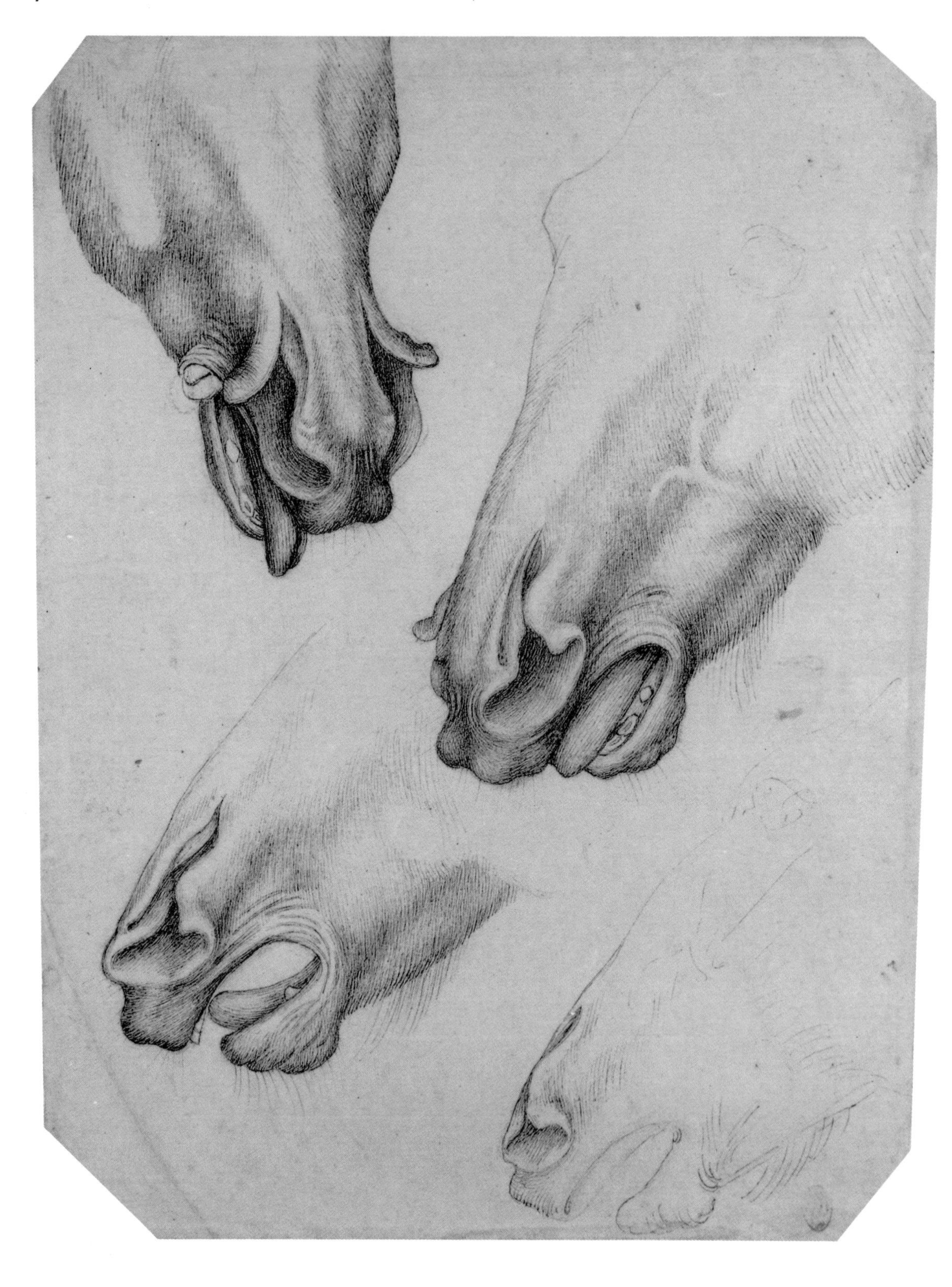

ig. 5.15 *Below*, a donkey with slit ostrils, at the present day. (Photo PANA.)

ig. 5.16 *Opposite*, horses with slit ostrils drawn by Pisanello. (Louvre, 'aris, photo Musées Nationaux.)

from Amarna, dated to about 1350 BC, and it is still being practised today as can be seen in the photograph of a living donkey (Fig. 5.15).

Figure 5.16 shows drawings that were made of horses with slit nostrils by the Italian artist Pisanello (AD 1395–1455), while from 200 years later the following graphic description of the operation is given in the seventeenth century account of a Portugese traveller in Arabia (Ley 1965):

> We can only have gone a short way after leaving the Arabs when the *shauter's* mare fell down suddenly as if dead. He made haste at once to cut its nostrils and cut something like an acorn out of the tear-ducts to its eyes. He then threw ground salt into its eyes and nostrils. After this it recovered again and could go on its way.

Such 'kill or cure' remedies have always pervaded veterinary practices as they have human medicine.

6 Equids in ancient Egypt and western Asia, and the enigma of the onager

The art of both the ancient Egyptians and the ancient Assyrians is perhaps best exemplified by the elegant and lifelike impressions of the chariot horses that are so well known from wall paintings and stone friezes. These horses have all the looks of the modern Arab with its small head, slender body, fine limbs, and high carriage of the tail (Fig. 6.1 & 6.2). With ostrich plumes on their bridles, bronze cheek-pieces, tassels on the harness, and groomed manes and tails, the horses fully enhance the splendour of the royal charioteers. But were these horses really the highly-bred stallions that they were made out to be? Figures 6.3 & 6.4 show two very similar ponies of the present day in Ethiopia, and it can be seen that the grey pony, decked out in festive regalia, immediately gives the impression of being much closer to an Arab type than the working pony with its handmade leather harness. This is because the neck of the grey is held in an arch by a tight rein and the mouth carries a bit and chain which inflict considerable discomfort if not actual pain, and these constraints give the pony a look of alertness.

The only certain way of knowing what the horses of the past looked like is from measurements of their skeletons, and because so few have been excavated from ancient Egypt they can be described here in some detail. The earliest skeleton is of a stallion, or possibly a gelding, that died in a fire which destroyed the fortress of Buhen in about 1675 BC. The remains of this fortress are near the second cataract of the Nile in northern Sudan. It was built during the early part of the Middle Kingdom (2053–1786 BC) and was one of a series of trading posts and strongholds erected for strategic purposes. The fortress was repeatedly attacked but it was during the first sacking that the horse, which had perhaps been stalled between two bastions of the main fortress wall, was killed and fell to the ground where its body lay on a brick pavement. It was covered with rubble and later the remains of the horse were sealed into this deposit by brickwork of the New Kingdom reconstruction of the fortress. The bones, which belong to the Khartoum Museum, could not be radiocarbon dated because all the collagen in them was degraded but the stratigraphic position of the skeleton establishes its date as *c.* 1675 BC.

ig 6.1 Tutankhamun 1361–1352 BC) in his chariot with ichly caparisoned horses, embossed n the gold base of the ostrich feather an. (Photo Robert Harding Picture ibrary.)

ig. 6.2 The Assyrian king ashurbanipal (668–627 BC) in his hariot on a stone relief from Nineveh. (Photo reproduced by ourtesy of the Trustees of the British Museum.)

Fig. 5.13 Scythian horsemen, apparently riding with stirrups, portrayed on the finials of a gold torque. Fourth century BC. Excavated from the Crimea in 1830. (Hermitage Museum, Photo Lee Boltin Picture Library.)

Fig. 6.3 Ethiopian pony, at the present day, with festive harness. (Photo author.)

Fig. 6.4 Working pony in Ethiopia at the present day. (Photo author.)

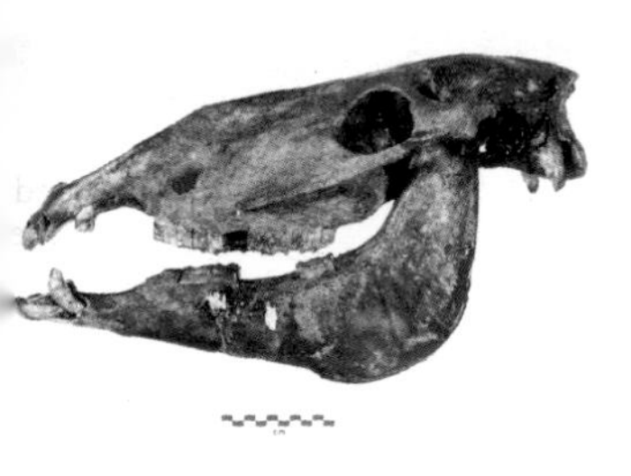

Fig. 6.5 The skull of the earliest horse to be excavated from the ancient Egyptian empire. From the fortress of Buhen, *c.* 1675 BC. *Top*, side view of the skull and mandible; *above*, enlarged view of the mandible to show the abnormal wear on the second premolar, probably caused by the horse chewing on a bit. (Photo The Natural History Museum, London.)

This nineteen year-old equid must have been a rare and valuable chariot horse. Not only is it the earliest horse to have been found in Egypt, but it also provides early material evidence for the use of a bit. This can be deduced from the excessive wear on the first lower cheek teeth (second premolars) which was most probably caused by continuous abrasion from a bronze bit (Fig. 6.5 Clutton-Brock 1974). The use of a bit at this early period is surprising as the pictures of horses in ancient Egypt usually show the bridle as having a nose-band, but no bit, as in Figure 6.1.

Two other horse skeletons have been described from ancient Egypt; one was a stallion from the Eighteenth Dynasty necropolis of Soleb dated to 1580–1350 BC (Ducos 1971) and the second was a mare that had been buried with a saddle cloth on its back and in its own coffin near the tomb of Sen Mut at Thebes in *c.* 1430–1400 BC (Boessneck 1970). Because the horses harnessed to chariots in the art of ancient Egypt and western Asia were always portrayed as stallions, it therefore seems likely that the mare buried at Thebes was a much valued riding horse, especially as she was interred with a saddle cloth.

The horse from Buhen was the largest; it was relatively fine-limbed and had a shoulder height of around 150 cm (14.5 hands). The Soleb equid was a pony that stood around 136 cm (13.5 hands) at the withers while the Thebes mare was rather taller at 143 cm (14 hands).

A fourth skeleton was excavated by Sir Flinders Petrie at the beginning of this century and presented by him to the British Museum (Natural History) but when it was dated by radiocarbon it was found to be less than four hundred years old, the date being 328±52 BP* (AD 1622) [BM-1357*] (Clutton-Brock & Burleigh 1979).

Finally, a skull and foot bones of a horse from Tell el Ajjul, Gaza, Sinai, which was also excavated by Petrie, has provided a radiocarbon date of 3400±120 BP [OxA-565], placing it, as Petrie believed, in the Hyksos period.

Both the horse from Buhen and the mare from Thebes probably looked much like the fine-limbed ponies to be found in Egypt and Arabia today, as did the seventeenth-century horse, but the pony from Soleb was probably more stocky. All four skeletons share a character in common with most Arab horses of today in that they had only five lumbar vertebrae. It is generally claimed that the Arab horse and the Przewalski horse have five lumbar vertebrae while all other breeds of horses have six. In fact the number is very variable but it is true that the Arab is more likely to have only five lumbar vertebrae than other breeds of domestic horse (Stecher 1962).

There are no fossil records of wild horse from North Africa after the

* See footnotes on page 58

Fig. 6.6 Skeletons of a pair of equids from Grave 168 at the Sumerian site of Abu Salabikh, Iraq. Excavated in 1981. (Photo N. Postgate.)

end of the Pleistocene period, so it is certain that the domestic horse was brought into the continent, and it is the Hyksos who are usually credited with the first imports. The Hyksos (meaning 'princes of foreign countries') first seized power in the region of the Nile Delta in about 1720 BC and then moved into the rest of Egypt. There were six Hyksos kings, and scarabs bearing their names have been found as far south as Sudan and in Palestine. The Hyksos set up trade contacts throughout Anatolia and western Asia, and they most probably brought horses and chariots into Egypt for use in royal processions and for hunting. The Hyksos rulers were ousted in about 1650 BC by a new family of kings (the Seventeenth Dynasty) who inherited the royal chariots. These became increasingly elegant and elaborate as can be seen from the chariot and horses that are engraved on the gold base of Tutankhamun's ostrich feather fan (1361–1352 BC, see Fig. 6.1).

As described in Chapter 5, the horse chariot was derived from simple carts and sledges that were originally yoked to cattle and only later to domestic asses. From graves at Ur in Mesopotamia (Fig. 4.5), dating to around 2500 BC, excavated by Sir Leonard Woolley in the 1930s, a number of four-wheeled vehicles were discovered including the famous sledge in the tomb of Queen Puabi (Shubad). At first Woolley believed that the skeletons of draught animals found with these vehicles were onagers, but they were later identified as cattle (Zarins 1986). Within the next hundred years or so the domestic donkey took over from cattle as the primary pack and draught animal and the two-wheeled chariot replaced the heavy four-wheeled cart. No remains of two-wheeled chariots have been excavated from this period, but there are a few clay as well as copper models of two-wheeled carts (Fig. 0.3). The ritual burial of two-wheeled carts or chariots can be inferred from the remains of pairs of asses with traces of harness which have been found in a number of Sumerian graves (Fig. 6.6; Clutton-Brock 1986, Zarins 1986).

Discrimination between the different species of equids on the fragmentary remains of their skeletons excavated from Middle Eastern sites dating to the third millennium BC is exceedingly difficult because, as known from the textual evidence as well as the osteology, asses, onagers, and horses were all cross-bred. However, it seems most probable that the equids buried in pairs in Sumerian graves were asses or hybrids between asses and onagers.

The replacement of the ass and its hybrids by the horse was probably a gradual process, but by the middle of the second millennium BC the light, fast, horse-drawn chariot with spoked wheels had become widespread in Egypt and western Asia. Asses and mules were to remain as pack and draught animals but the riding of horses was beginning. In the earliest portrayals the rider is seated as if riding an ass, that is well back on the loins of the horse (Littauer & Crouwel 1979; Fig. 4.13). Because the ass has very low withers the rider must sit far back to avoid sliding forward over the animal's neck, whereas the horse rider sits forward and is balanced by the high point of the withers. It is probable that as the horse became a more accepted mode of transport its greater speed was appreciated, but it was slow to supplant the ass and the mule and it is not until the eighth century BC that Assyrian riders are shown sitting at their ease, bareback behind the withers with their legs hanging naturally (Fig. 6.7; Littauer & Crouwel 1979).

There is no evidence for the ass being used to draw a chariot in ancient Egypt but hybrids can be seen in the well-known pair of wall

Fig. 6.7 King Ashurbanipal shooting at wild onagers (compare with Figure 5.11). Stone relief from Nineveh, *c.* 645 BC. (Photo reproduced by courtesy of the Trustees of the British Museum.)

Fig. 6.8 Horses (above) and hinnie (below) harnessed to chariots on a tomb painting from Thebes, Egypt, *c.* 1400 BC. (Photo reproduced by courtesy of the Trustees of the British Museum.)

paintings from a tomb at Thebes, dated to around 1400 BC (Fig. 6.8). The upper picture shows a pair of typical chariot horses but the lower one is of two very unusual equids. They have been variously interpreted as asses or onagers, but their physical appearance perhaps most closely resembles that of male hinnies, of which a living example is shown in Figure 3.5. A hinny (as described in Chapter 3) is the offspring of a male horse and a female donkey; it is incapable of breeding but its external reproductive organs are fully developed as can be seen in the Egyptian painting. This pair may have been held in very high esteem because they were white. Another pair of hinnies harnessed to a chariot is portrayed in a harvest scene of the Eighteenth Dynasty (1567–1320 BC) from Thebes (Fig. 6.9).

Fig. 6.9 A pair of hinnies in a harvest scene from the Tomb of Khaemhet, Thebes, Dynasty XVIII, 1567–1320 BC. (Photography by the Egyptian Expedition, Metropolitan Museum of Art.)

The enigma of the onager, *Equus hemionus*. Was it ever domesticated?

It was the site of Tell Asmar that provided the apparent osteological evidence for the theory that the onager was the domestic equid of ancient Mesopotamia (Hilzheimer 1941; Fig. 4.5). Because such large numbers of hemione remains were identified from this Sumerian site, Zeuner ignored the probability that these were wild animals killed for their meat and assumed them to be domestic. He wrote in his classic *A History of Domesticated Animals* (1963): 'It is not generally known that, prior to the introduction of the domesticated horse into Mesopotamia, the half-ass or hemione was domesticated by the ancient Sumerians.'

Zeuner was of course familiar with the much later Assyrian stone reliefs from the palace of Nineveh where onagers are shown being hunted and caught, in the wild, with ropes (Fig. 6.10), and he considered that these animals may have been then tamed and used to draw chariots. Zeuner did also discuss the probability that onagers were interbred with domestic donkeys, which recent textual evidence indicates was the only purpose for which they were kept in captivity. Following Zeuner's writings it was generally believed that the onager was the domestic equid of the ancient Near East, and that it ceased to be used in the second millennium BC when it was supplanted by the horse. However, if the onager had been used as a domestic ass throughout western Asia it could be expected that the

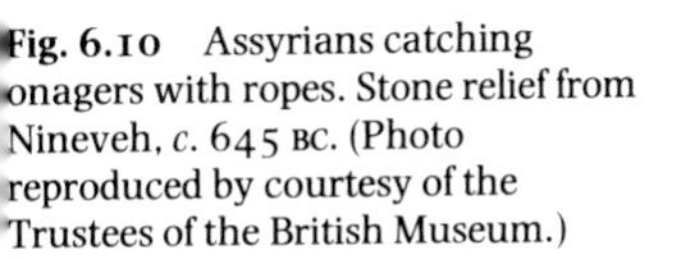

Fig. 6.10 Assyrians catching onagers with ropes. Stone relief from Nineveh, *c.* 645 BC. (Photo reproduced by courtesy of the Trustees of the British Museum.)

Fig. 6.11 Victory scene on the wooden box known as the Standard of Ur. (Photo reproduced by courtesy of the Trustees of the British Museum.)

modern domestic asses of the region today would be descended from this ancient stock, but this is not so. As far as is known all the donkeys of Asia are descended from the African wild ass, *Equus africanus* and not from the onager, *Equus hemionus*. This can be deduced from the fact that the donkey will only produce fertile offspring when it is bred with the African wild ass, and not when it is crossed with the onager (Gray 1971).

The pictorial evidence for the domestication of the onager was supposedly provided by the 'Standard of Ur', a box measuring approximately 48.3cm by 20.3cm, covered with elaborate pictures in inlay and dating to around 2500 BC (Fig. 6.11). This famous box, which was excavated from a grave at Ur (Fig. 4.5), portrays on one side a battle scene and on the other what may be presumed to be the ensuing victory. Zeuner reproduced photographs of these scenes (1963) with the comment: 'The first and third registers of the War Panel show four-wheeled chariot drawn by onagers (note the tails!).' However, if he had also noted the shoulders of the equids in these little pictures he would have seen that they are marked with a strong shoulder stripe which is diagnostic of the donkey (*Equus asinus*), and is not present in the onager.

Another object which has been taken as evidence of a domestic onager is the figure of an equid in electrum (an alloy of gold and silver) on a double rein ring found in the grave of Queen Puabi (Shub-ad), also at Ur (Fig. 6.12). This probably does represent an onager, but it should not be assumed that it was domesticated for rein rings were very often decorated with models of wild animals such as a lion, stag, or ibex.

Fig. 6.12 Onager on the rein ring from the grave of Queen Puabi at Ur, Iraq, *c.* 2500 BC. See also Figure 5.3. (Photo reproduced by courtesy of the Trustees of the British Museum.)

Textual evidence for equids in third millennium BC Mesopotamia

The Sumerians of ancient Mesopotamia were the first to undertake the breeding of equid hybrids and to document what they did. This was done in cuneiform script written on clay tablets. The Sumerians, like the ancient Egyptians, were great believers in hierarchy and order, and they made lists of everything with which they were concerned in their daily lives. Decipherment of the cuneiform has itself a long history. The first great collection of over 20 000 clay tablets was discovered, in the 1840s, by Austen Henry Layard in the royal library of King Ashurbanipal (668–627 BC) in the palace at Nineveh in Assyria (Fig. 4.5). Well before this, in 1802, Grotefend, a German scholar, had made efforts to translate tablets that had been discovered, written in the Old Persian script, but it was not until 1847–53 that an English army officer, H. C. Rawlinson, amongst other scholars, managed to decipher inscriptions in the far more complicated Babylonian version of this script. By the end of the nineteenth century the decipherment of the tablets from King Ashurbanipal's library was underway but scholars today still continue with the translation and interpretation of innumerable clay tablets and stone inscriptions from sites throughout Mesopotamia. 'Assyria' (the land of Assur or Ashur as it is also spelt) is the name given to the empire of northern Mesopotamia which dated from 2000 BC until it fell in 612 BC. Babylonia is the name given to southern Mesopotamia, the land of the earlier Sumerian Empire which lasted throughout the third millennium BC.

Within recent years scholars have given diligent attention to sorting out the terms, written in cuneiform, that the Sumerians used for their different equids (Zarins 1978; Maekawa 1979*a*, 1979*b*; Maekawa & Yildiz 1982, Postgate 1986). The terms can now be summarized, following Postgate (1986):

anše* = generic term for any equid or domestic ass (*E. asinus*)
anše-DUN.GI or anše-LIBIR = domestic ass (*E. asinus*)
anše-eden-na = onager (*E. hemionus*)
anše-BARxAN = hybrid *E. asinus* × E. hemionus
anše-zi-zi or anše-kur-ra = domestic horse (*E. caballus*)

*In quoting words from cuneiform texts there is a standard convention that Roman letters are used for Sumerian, eg. anše, and capitals are used for signs of which the phonetic equivalent is not known, even though the meaning may be certain eg. LIBIR.

It seems clear from the texts that the domestic donkey was the common agricultural draught animal, while small numbers of onagers were kept for cross-breeding with donkeys. The hybrid offspring were large powerful animals which were yoked in teams and were used to draw chariots. They would have been quite as large as the early domestic horses. Onagers were also hunted as wild animals and may have been valued for their meat and hides but there is no evidence that they were used directly as draught animals. Thus the evidence from the texts fully corroborates the identification of the osteological remains from graves as ass and ass/onager hybrids rather than onagers.

The horse is mentioned occasionally in the cuneiform scripts from around 2050 BC. One three-year account tablet records 37 horses (anše-zi-zi), 360 onagers, 727 hybrids, and 2204 donkeys (Postgate 1986). The horse remained a high status animal, but it is mentioned increasingly often in the texts and at the same time the onager disappeared completely. To quote from Postgate (1986):

> Presumably once the horse, which could not only breed hybrids but was also in itself a better animal and trainable, had appeared on the scene, there was no longer any call for keeping the onagers to do nothing but eat, drink, and enjoy the pleasures of their harems. The word which had referred to the donkey × onager hybrid did survive, though, because it was simply redeployed for the new donkey × horse hybrid.

The Assyrian palace reliefs

> The assyrian came down like the wolf on the fold,
> And his cohorts were gleaming in purple and gold;
> And the sheen of their spears was like stars on the sea,
> When the blue wave rolls nightly on deep Galilee.
>
> And the widows of Ashur are loud in their wail,
> And the idols are broke in the temple of Baal;
> And the might of the Gentile, unsmote by the sword,
> Hath melted like snow in the glance of the Lord!

Lord Byron (1788–1824) was able to write *The destruction of Sennacherib* (of which the above are the first and last verses), because his classical and biblical education enabled him to know much about the might of the ancient Assyrian empire of the first millennium BC. There had, however, been no excavation of the ruined palaces of Nimrud and Nineveh during his lifetime, so Byron had no knowledge of the great carvings and stone reliefs that were first made known to the western world in the middle of the nineteenth century.

The ancient palaces of Assyria were built as a series of rectangular courts surrounded by long rooms, the stone walls of which were decorated with elaborately carved bas-reliefs (sculptures in low relief). Almost all the kings built a new palace, and their ruins now cover a huge area which has been subjected to archaeological excavation intermittently over the last hundred and fifty years.

The most numerous and grandest of the stone reliefs come from the palace of Ashurnasirpal II (883–859 BC) who moved the capital of Assyria from Assur to Nimrud (Fig. 4.5) where it remained until Sennacherib (704–681 BC) moved it again to Nineveh. Nimrud was named by the Arabs in honour of the Biblical hero Nimrod, 'mighty hunter before the Lord', but it was known as Kalhu to the Assyrians and appears in the Old Testament as Calah. Stone reliefs in huge numbers decorated the walls of the palaces at Nimrud and at Nineveh, and the carving of scenes of hunting and war, particularly those from the palace of Ashurbanipal (668–627 BC) at Nineveh, have a forcefulness and realism that is only matched in genius by the classical Greek sculptures.

These later stone reliefs, many of which are held in the British Museum, depict in marvellous detail the pageantry of the nobility and the suffering of the conquered. What greater magnificence could be portrayed than that of King Ashurbanipal in his triumph over Shamash-shumm-ukin as he stands in his chariot, his horses bedecked with finery, his wife behind him, and an umbrella over him as a protection from the sun (Fig. 6.2).

The chief form of transport for the élite in battle and in the hunt was the horse chariot, but the riding of horses by the nobility for hunting and by ordinary soldiers in war was by this time well established, as can be seen in Figure 6.7. In Figure 6.13 carefully groomed horses with parted manes and braided tails are being led in procession. The horses are held by a rope that is tied tightly round the diastema of the lower jaw, that is between the canine teeth and the premolar cheek teeth. This method of control, which can be very cruel because it causes wounds to the jaw, was common throughout the ancient world and is still sometimes used at the present day.

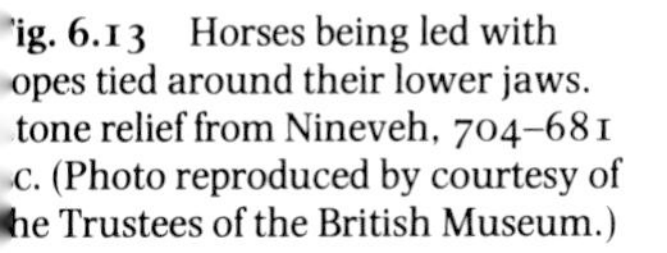

'ig. 6.13 Horses being led with opes tied around their lower jaws. tone relief from Nineveh, 704–681 c. (Photo reproduced by courtesy of he Trustees of the British Museum.)

Fig. 6.14 A mule carrying a woman and child. Stone relief from Nineveh, *c.* 645 BC. (Photo reproduced by courtesy of the Trustees of the British Museum.)

There are no depictions of the domestic donkey on the Assyrian stone reliefs which are in the British Museum, presumably because this equid had no place in either hunting or warfare, the main subjects of the friezes. However, the mule is shown and was clearly an important baggage animal; in Figure 6.14 a woman (perhaps the queen who had been taken prisoner) is riding a mule with her child and a man is walking behind carrying a bag of belongings. From their horse-like tails and general appearance, which is most realistic, these mules must have been hybrids between an ass and a horse and not between an ass and an onager.

The scene of wild onagers being caught with ropes, also from the palace of Ashurbanipal (Fig. 6.10), is something of an anomaly, because it implies that onagers were being kept in captivity for some purpose. This goes against the textual evidence as interpreted by Postgate (see above), who claims that the cuneiform sign for the onager disappears from the later texts. Perhaps wild onagers were caught so that they could be transferred to the scene of a royal hunt, for on another relief King Ashurbanipal is shown as an archer on horseback shooting arrows into fleeing onagers, likewise a lion is shown being let out of a cage so that it could be killed by the king.

Horses and asses in the Old Testament

THE HORSE

Hast thou given the horse strength? has thou clothed his neck with thunder? Canst thou make him afraid as a grasshopper? the glory of his nostrils is terrible. He paweth in the valley, and rejoiceth in his strength: he goeth on to meet the armed men. He mocketh at fear, and is not affrighted; neither turneth he back from the sword. The quiver rattleth against him, the glittering spear and the shield. He swalloweth the ground with fierceness and rage: neither believeth he that it is the sound of the trumpet. He saith among the trumpets, Ha ha; and he smelleth the battle afar off, the thunder of the captains, and the shouting. *Job* 39, 19–25

The main references to the horse, as well as to other animals in the Bible, were collated by Tristram (1889) who was not only an erudite Canon of the Church of England but also a naturalist with a wide knowledge of the fauna and flora of Palestine in the last century. He noted that throughout the Old Testament the horse is almost always described in connection with armies and battles, as in the above quotation. It is only very occasionally that the horse is mentioned in an agricultural context, which was the preserve of the ass.

Tristram gives four different Hebrew words for the horse, the most common being *soos* meaning a chariot horse. Then there is *recesh*, a swift or high-bred horse, although this word may also have meant a dromedary, as in, 'And he wrote in the king Ahasuerus' name and sealed it with the king's ring, and sent letters by posts on horse-back, and riders on mules, camels, and young dromedaries' (Esther 8, 10). *Rammac* means a mare and *parash* means a horseman, cavalry horse, or riding horse in contrast to a chariot horse, as in, 'And Solomon had forty thousand stalls of horses for his chariots and twelve thousand horsemen' (I Kings 4, 26).

The first mention of the horse in the Bible is when Joseph saved the people of Egypt from the great famine described in Genesis, 'and Joseph gave them bread in exchange for horses, and for the flocks, and for the cattle of the herds, and for the asses' (Genesis 47, 17). From this time onward, throughout the second and first millennia BC, the period of history covered by the Old Testament, the chariot horse played an integral part in the endless wars fought between the peoples of western Asia. The Israelites came to depend on a supply of horses and chariots from Egypt to defend themselves against Syria and the Assyrians, a policy which was denounced by the prophets, 'Woe to them that go down to Egypt for help; and stay on horses, and trust in chariots, because they are many; and in horsemen because they are very strong;' (Isaiah 31, 1); 'Now the Egyptians are men, and not God; and their horses flesh, and not spirit' (Isaiah 31, 3).

This was the period during which iron was coming increasingly into use throughout Europe and Asia, and the consequent improvements in the design of chariots is noted in, 'And the Lord was with Judah; and he drave out the inhabitants of the mountain; but could not drive out the inhabitants of the valley, because they had chariots of iron' (Judges 1, 19). The chariots were not of course made entirely of iron and what was meant was that iron tyres were nailed onto the wooden wheels.

THE DOMESTIC ASS

Tristram (1889) commented that, "The ass is frequently mentioned in Scripture as being ridden by persons of wealth and quality, as indeed it is to the present day in the East'. Where the horse provided the means of transport in war, the domestic ass provided it in peace and on the farm. Women and children are frequently recorded as riding on an ass as in, 'And Moses took his wife and his sons, and set them upon an ass, and he returned to the land of Egypt' (Exodus 4,20).

White asses were particularly favoured and owned by persons of high rank, as shown by Deborah in her exhortation to the judges, 'Speak ye that ride on white asses, ye that sit in judgement, and walk by the way' (Judges 5, 10). In the famous wall painting from Beni Hasan in Egypt (*c.* 1900 BC) a Semite is shown with his white ass (Fig. 0.2).

To own a large number of asses conferred high status as for the judge Abdon, 'And after him Abdon the son of Hillel, a Pirathonite, judged Israel. And he had forty sons and thirty nephews, that rode on three score and ten ass colts; and he judged Israel eight years' (Judges 12, 13–14).

The ass was commonly used for ploughing: 'The oxen likewise and the young asses that ear [till] the ground shall eat clean provender, which hath been winnowed with the shovel and with the fan' (Isaiah 30, 24). It was, however, forbidden to yoke the ox and the ass together, 'Thou shalt not plough with an ox and an ass together' (Deuteronomy 22, 10).

THE MULE

Tristram (1889) gives the Hebrew words *pered* and *pirdah* for the mule, and assumes that this was the hybrid donkey × horse, but it is possible that the onager × donkey was also sometimes inferred.

Mules had a high status in Old Testament writings as shown in the First Book of Kings where Solomon was described as riding in state on a mule when he was declared king:

> The king also said unto them, take with you the servants of your lord, and cause Solomon my son to ride upon mine own mule, and bring him down to Gihon: and let Zadok the priest and Nathan the prophet anoint him there king over Israel: and blow ye with the trumpet, and say, God save King Solomon.
>
> *I Kings* 1, 33–34

The mule was also used as a baggage animal but it seems probable that the Israelites, at least in the early period, were forbidden to breed mules, for as Tristram notes, it was a law of Moses that, 'Thou shalt not let thy cattle gender with a diverse kind' (Leviticus 19, 19).

THE ONAGER

References to the *pere* in the Old Testament have been translated as 'wild ass', but it is the onager (*Equus hemionus*) that is being described. Perhaps the most evocative passage has been quoted on p. 38, but many other allusions to the wild and untameable nature of the onager show the respect in which nature was held in the ancient world, 'A wild ass used to the wilderness, that snuffeth up the wind at her pleasure; in her occasion who can turn her away?' (Jeremiah 2, 24).

Tristram (1889) noted that there are two references in Job (39, 5) and in Daniel (5, 21) to the Hebrew *Arôd*, which is also translated as 'wild ass', and he questions whether it is the African wild ass (*Equus africanus*) that is here intended. If so it would lend support to evidence from the archaeozoological record, for a small number of equid bones from early prehistoric sites in Arabia has been identified as *Equus africanus* (Uerpmann 1991).

7 The horses of Scythia and the Orient

The steppe nomads of Eurasia had no rivals as horsemen, from the first millennium BC until the great empire of Genghis Khan in the middle ages. In the fifth century BC Herodotus wrote about the Scythians and other mysterious tribes who inhabited the land, '... to the northward of the furthest dwellers in Scythia, [where] the country is said to be concealed from sight and made impassable by reason of the feathers [snow] which are shed abroad abundantly.' (IV, 7, Rawlinson 1964). The land of these nomadic pastoralists covered a vast area from directly north of the Black Sea eastwards and north to Mongolia and southern Siberia (Fig. 4.3).

In order to understand the way of life of these people, some of whose descendants still live as pastoralists in the Altai mountains (Vainshtein 1980), it is necessary to give a brief summary of the origins of nomadic pastoralism and the terms used. Varro, the Roman writer on agriculture, following the ancient Greeks, believed that, 'in the primitive period, when people led a pastoral life, they were ignorant even of ploughing, of planting trees, and of pruning, and that agriculture was adopted by them only at a later period' (I,ii,16, Hooper & Ash 1967). Since the time of Varro, historians have believed that primitive hunters became pastoralists, that is herders of grazing animals, before they became farmers, and the reindeer herders of the north were taken as an example of how this could have happened. Hunters who followed herds of wild reindeer were presumed to have learned how to control the living animals which then gradually became domesticated.

It would follow from this hypothesis that settlement and the beginnings of agriculture developed later than pastoralism. All the evidence now indicates, however, that nomadic pastoralism in Eurasia was an adaptive strategy adopted by agricultural communities living in regions of hardship where the climate was becoming increasingly arid (Khazanov 1983).

Fig. 7.1 Plaited horses' tails from the barrows at Pazyryk, fifth centur BC. From Rudenko (1970).

The main terms used in studies of pastoralism should be briefly defined here. *Settled agriculturalists* live by cultivating crops and husbanding livestock which may be herded over short distances for grazing but they always return to the home base; *transhumance* is the term given to the regular movement of flocks and herds by *pastoralists* between mountain and lowland pastures at different seasons of the year; *nomadic pastoralists* have no home base and do not grow crops, although they may obtain grain and vegetables by trade. Nomadic pastoralists live by following their herds to optimum grazing lands at different seasons of the year; they may migrate over large distances, and they subsist almost entirely on the resources provided by their animals: meat, milk, blood, and hides.

Scythian nomads in the first millennium BC

From the fourth millennium BC domestic horses were bred in small numbers on the steppes of central Asia, along with cattle and sheep, for their meat but, as the harsh environment became ever more arid, treeless and over-grazed, horses became the predominant livestock. There is always a tendency amongst pastoralists to increase the numbers of their animals beyond the carrying capacity of the land, and on the steppes the only way to manage the proliferating herds of horses was by riders who could keep up with their migrations. Thus, by the first millennium BC, horse-riding had become the established means of locomotion in central Asia, and it enabled nomadic pastoralism to develop into an elaborate cultural economy.

These nomads who ranged over the whole of central Eurasia are better known to us than any other illiterate society of the period. Although the Scythians had no written language the powerful union of tribes that lived in the region were described by the Greek writers, notably Herodotus and Strabo, and by inference Homer (Iliad xiii, 1–7), all of whom were awed by their wandering way of life. Furthermore, the famous gold treasure obtained from Siberia by Peter the Great in the eighteenth century, and the remarkable discoveries from the frozen tombs of Pazyryk in the Altai mountains (Fig. 4.3) corroborate much of the Greek writings and reveal a wealth of information about the earliest nomadic horsemen (Figs. 7.1–7.4).

g. 7.2 Mask for a horse's head ecorated with a wooden stag's head nd antlers made of leather from azyryk. From Rudenko (1970).

While the Assyrians and the ancient Egyptians were building massive temples and perfecting the horse chariot for warfare and hunting, the Scythians were developing into an élite society of nomads who had no settled places except for the tombs where they buried their dead. They were for long periods peaceful people except when their sacred tombs were interfered with, as recorded by Herodotus in the reply of the Scythian king Idanthyrsus when invited to engage in battle by Darius the Persian:

> We Scythians have neither towns nor cultivated lands, which might induce us, through fear of their being taken or ravaged, to be in any hurry to fight with you. If, however, you must needs come to blows with us speedily, look you now, there are our father's tombs – seek them out, and attempt to meddle with them – then ye shall see whether or no we will fight with you. *IV, 127, Rawlinson 1964*

Fig. 7.3 The four-wheeled covered carriage from Barrow 5, Pazyryk. From Rudenko (1970).

The Scythians were by no means always peaceful, however, and where their territories impinged on those of other cultures they were very ready to fight, and indeed they contributed to the fall of the Assyrian empire in the late seventh century BC.

The Scythians lived in movable felt tents and owned many elegant and sophisticated, yet portable, belongings. They had trading links throughout the Near East as well as with China and had an abundance of gold which they used to ornament their beautifully made saddlery, carpets, and clothes. Horses were their most highly valued possessions and were therefore essential victims of sacrifice, as so lucidly recorded by Herodotus in his description of the final burial of a king:

> When a year is gone by, further ceremonies take place. Fifty of the best of the late king's attendants are taken, all native Scythians – for as bought slaves are unknown in the country, the Scythian kings choose any of their subjects that they like, to wait on them – fifty of these are taken and strangled, with fifty of the most beautiful horses. When they are dead, their bowels are taken out, and the cavity cleaned, filled full of chaff, and straightway sewn up again. This done, a number of posts are driven into the ground, in sets of two pairs each, and on every pair half the felly* of a wheel is placed archwise; then strong stakes are run lengthwise through the bodies of the horses from tail to neck, and they are mounted up upon the fellies, so that the felly in front supports the shoulders of the horse,

Fig. 7.4 The wall hanging from Barrow 5, Pazyryk. The decoration made from coloured felt, c. 430 BC. From Rudenko (1970).

Fig. 7.8 Modern bridle on a horse Tuva, southern Siberia, to compare with those shown in Figures 7.4 a 7.5. (Photo C. Humphrey.)

Fig. 7.12 Bronze statuette of a Ferghana horse with its foot on a swallow, first century AD. Found in tomb at Lei Tai in the Kansu provin of China in 1969. (Photo Robert Harding Picture Library.)

* A felly or felloe is the rim of a whe

7.4

7.8

> while that behind sustains the belly and quarters, the legs dangling in mid-air; each horse is furnished with a bit and bridle, which latter is stetched out in front of the horse, and fastened to a peg. The fifty strangled youths are then mounted severally on the fifty horses. *IV,72, Rawlinson 1964*

Although they are several thousand kilometres away from the region of the Caucasus that was known as Scythia to Herodotus, the frozen tombs of Pazyryk in the Altai mountains have shown that many of his accounts were correct. These tombs which, like the history of Herodotus, date to around the fifth century BC, have been known about for hundreds of years and much of the wonderful gold treasure and textiles, which had been preserved, frozen in permanent ice, can be seen today in museums and collections in Russia and Asia. Unlike most of the tombs which were opened in antiquity, the horse burials were usually left intact until they were excavated, between 1929 and 1949, by the Russian anthropologist, Rudenko. Five large tombs and a number of small ones, which have been shown by tree-ring analysis to span a period of about fifty years, were excavated. Each of the large tombs contained the remains of between seven and fourteen horses, there being a total of at least fifty-four skeletons.

Fig. 7.5 A Scythian bridle from Pazyryk, fifth century BC. After Rudenko (1970).

As described by Rudenko (1970) all the large tombs were of the same construction: each was covered by a low earthen mound and a cairn of rocks. Underneath this was a rectangular shaft, facing east-west, with a human burial in a wooden-lined chamber on the south side. Outside the chamber, in the northern part of the shaft, there were horse burials. The horses had been killed, not by strangling as described by Herodotus, but by pole-axing, that is by a heavy blow on the front of the head. In nearly all cases the head of the horse faced to the east. With the exception of Tomb 5 where there were four draught horses, their bridles, and a four-wheeled carriage, all the rest were riding horses buried with saddles, bridles, bronze or iron bits, and whips; and some even had very elaborate felt masks and head-dresses still in place on the skull (Figs. 7.5–7.7). The horse's ears had been clipped with ownership-marks, their manes were clipped and covered with leather, and their tails were plaited or knotted.

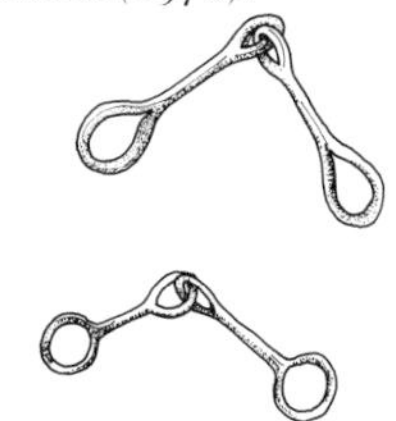

Fig. 7.6 Scythian bronze bits from Pazyryk, fifth century BC. After Rudenko (1970).

In most of the large tombs there were the remains of wooden carts or trolleys with solid wooden wheels, but in Tomb 5 there was a very unusual four-wheeled carriage with draught pole and four harnessed horses, two with yoke saddles and reins, and two with traces (Fig. 7.3). Rudenko (1970) considered that this carriage, which had wheels with multiple spokes and a felt canopy decorated with embroidered swans, was of Chinese origin.

Fig. 7.7 A horse's head-dress made felt and topped with a goat's head and a bird, also made of felt, with leather trimmings, from Barrow 2 at Pazyryk, fifth century BC. After Rudenko (1970).

From all these remains it has been possible to build up a very detailed picture not only of Scythian horse-gear, but also of the kind of horses and how they were fed and stabled in the fifth century BC.

The bones of the horses have been measured and analysed by Vitt (1952) and by Bökönyi (1968) who both agreed that there were two kinds of horses in the tombs. Those with the masks were clearly the most highly valued horses; their mummified coats were still obviously sleek and they were the tallest with withers heights of around 14 hands 1 inch (140–145 cm). The smallest horses were only about 13 hands (132 cm) at the withers and were more stocky. Vitt suggested that the taller horses may have been gelded as foals for if animals are castrated before they are full grown their bones will increase in length so that when adult they are taller than either the male or female, and the skull, particularly in the facial region, also becomes longer, giving the horse a more 'purebred' look (Littauer 1971). Rudenko (1970) stated that, 'In the tombs of noble persons at Pazyryk all the horses were geldings; not a single stallion or mare was found.' However there is no further documentation about precisely which horses were connected with the 'noble persons'. That the castration of horses was an unusual practice in the classical world was recorded by Strabo who wrote, sometime after 44 BC:

> It is a peculiarity of the whole Scythian and Sarmatian race that they castrate their horses to make them easy to manage; for although the horses are small, they are exceedingly quick and hard to manage.
>
> *7, 4, 8, Jones 1983*

Like Herodotus, writing four hundred years earlier, Strabo obviously found the customs of Eurasian nomads strange for he remarks a number of times that:

> In fact, even now there are Wagon-dwellers and Nomads, so-called, who live off their herds, and on milk and cheese, and particularly on cheese made from mare's milk and know nothing about storing up food or about peddling merchandise either, except the exchange of wares for wares.
>
> *7, 3, 7, Jones 1983*

This statement could equally well have been made at the beginning of this century, for the nomads of the Eurasian-mountain steppes have had a remarkable history of continuity over at least three thousand years, as shown by Vainshtein (1980) in his study of pastoralist society in the isolated mountain province of Tuva (Fig. 4.3). Right up to the nineteenth century a horse was killed and buried with its Tuvinian rider, and it can be seen in Figures 7.5 & 7.8 that the bridle in use today in southern Siberia is very similar in design to those from the Pazyryk tombs. Mares were milked from between five to seven times a day because they produced milk rapidly, according to Vainshtein, and milking was carried out from June to September with a daily yield of around 2.5 litres. By ancient tradition, mares were milked by men, while other animals were milked by women.

With one exception, there is no evidence that the Scythians ever rode with stirrups. No stirrups have been found in the Pazyryk tombs and they are not mentioned in any Greek or Roman writing, nor are they seen in any classical art. However, as discussed in Chapter 5 (p. 76) and shown in Figure 5.13, there is one depiction of a metal hook stirrup on a gold torque of Scythian date. This seemingly simple aid to staying on the back of a horse did not become widely used until the post-Roman period.

Equids in the Orient

There may have been domestic horses in China as early as the Luangshan period which dates from 3000–2300 BC (Chow 1989). If the few remains of equids from sites of this period are really attributable to domestic horses, this would accord well with the evidence from the Ukraine for the breeding of horses, as early as 4000 BC (see Chapter 4). However, it was not until the Bronze Age, nearly a thousand years later, that the horse became of crucial importance for transport and warfare throughout China.

Fig. 7.9 The head of one of the full size terracotta horses from the mausoleum of the Emperor Qin Shihuang near Xian in the Shaanxi province of China, 210 BC.

The Luangshan culture was followed by the Shang Dynasty, in which, after 1400 BC in the late Yin phase, the two-wheeled chariot makes its first appearance in China. Remains of nearly a hundred horses from sacrificial pits at Yin Hsu have been studied in detail by Chow (1989). The horses had withers heights of between 133 and 143 cm, heavy heads, and stocky bones, indicating that they resembled the modern Przewalski horse. The Chinese horses gradually increased in size, but they remained heavy in build until after the Qin Dynasty when Qin Shihuang, who died aged 49 in 210 BC, was buried in what was probably the most astonishing mausoleum ever to be constructed. The tomb has never been excavated but in three vast underground chambers there were buried an estimated 7500 life-size pottery figures of soldiers and horses which have become familiar from exhibitions worldwide of examples from this terracotta army (Fig. 7.9).

Qin Shihuang, the first Emperor of China, was born in 259 BC and came to the throne of the Qin Dynasty when he was only thirteen years old. His life and times are recorded in great detail in the *Shi Ji* (Records of the Grand Historian, quoted here from Wirgin 1985). At first the affairs of state were attended to by the Prime Minister Lu Buwei, a business man who built up great riches, but the young king gradually took control and it is obvious that he suffered from an extreme form of megalomania combined with an increasing and overwhelming terror of death. As was the custom, when he was quite young, Qin Shihuang began the planning and construction of

'ig. 7.10 A warrior with his horse rom the mausoleum of the Emperor)in Shihuang.

his tomb, using a labour force of seven hundred thousand men. According to the *Shi Ji* (Wirgin 1985):

> They dug through three subterranean streams and poured molten copper for the outer coffin, and the tomb was fitted with models of palaces, pavilions and offices, as well as fine vessels, precious stones and rarities.

All the king's wives who had not borne sons were interred with him, and all the artisans and labourers were imprisoned between the middle and outer gates. Trees and grass were planted over this vast tomb to make it look like a hill which is surrounded by a wall 2000 metres in length. The main tomb has not been excavated, at least in modern times, but in 1974, when drilling for wells, some local farmers discovered large fragments of pottery which led to the discovery of the now famous buried army. Three very large pits were found to contain 7500 life-size figures of soldiers and horses made out of fired clay, as well as the remains of wooden chariots. The pits are five to seven metres below the present ground level and the figures are placed in corridors with their backs to the main tomb (Wirgin 1985; Fig. 7.10).

In 1980, excavation between the inner and outer walls of the burial mound revealed two remarkable carriages buried at a depth of seven to eight metres. The carriages are made entirely of bronze and are in a very fine condition (Fig. 7.11); each is harnessed to four bronze horses which are of the same stocky build as the clay models. One carriage is larger than the other and is approximately half life-size. It is 2.86 metres long, 1.07 metres high and weighs 1241 kg; it is covered with a very thin canopy of cast bronze and was originally lined with silk. The carriages are presumably models of those used by the royal family, or perhaps for the final funeral service. Their manufacture had been carried out with the utmost precision and the one shown in Figure 7.11 is made up of 3462 parts of gold, silver, and bronze (Wirgin 1985). It is evident that, by the third century BC in China, the technology of casting metal had achieved a very high degree of sophistication, but the horses still had the heavy build of the modern Mongolian pony and the harness was still the same as that used throughout the Near East and the classical world. The four horses were harnessed as a quadriga, that is the inner two are yoked to a pole and the outer pair are held with traces. It is noteworthy that the horses' manes are clipped, their tails are braided, and the bridle has the weighted 'pom-pom' so often seen in the Assyrian stone reliefs.

A hundred years later, news of the first Ferghana heavenly horses was brought to the Western Han emperor by the great traveller Chang Chi'en. Ferghana was in Turkestan and was the eastern outpost of the classical world. The horses were larger and more fine-limbed than any seen before in China. They were said to 'sweat blood', which, as quoted by Epstein (1969) was due to a parasite, *Parafilaria multipapillosa*, which causes slight bleeding of the skin and so makes the foamy sweat pink.

In 104 and 103–101 BC, after the ruler of Ferghana had refused to barter thoroughbred horses, and the Chinese envoys sent to buy them had been murdered, Wu-ti, the emperor, sent two large expeditionary forces the 3000 km to obtain the horses by force. Over

Fig. 7.11 One of the bronze carriages from the mausoleum of the Emperor Qin Shihuang.

Fig. 7.13 Model of a Japanese warrior, representing the armour and weapons of the late sixteenth century AD. (Photo Victoria and Albert Museum.)

3000 horses were collected but only 50 'blood-sweating' and 1000 inferior horses survived the long desert journey back to China. The Ferghana horses, however, were invested with religious dignity; they were 'heavenly' and therefore to be used at the Emperor's funeral. They became a legend and were a focus for many of the most beautiful paintings and sculptures in Chinese art (Fig. 7.12 see p. 98).

The ass, mule, and hinny have had a long history in China; the ass was first introduced there in the third century BC and the mule has been the most valuable pack animal since ancient times. Very large mules, standing 16 hands (163 cm) at the withers are not uncommon and are usually bred from Mongolian pony mares. The hinny, that is the cross between a donkey jenny and a pony stallion, has been traditionally bred for drawing carts and even today is more commonly seen than in other countries.

In Japan, according to Naola (1970), remains of horses appear in archaeological sites belonging to the late Jomon, or Neolithic period, that is during the first millennium BC. Presumably these were domestic horses whose progenitors were brought to Japan in boats from China or Korea. As agricultural practice developed in Japan cattle became the animals for draught and there is no evidence that horses were ever driven (Littauer, pers. comm. 1991). During the Middle Ages horses were of great importance as essential accoutrements to the armed Samurai warriors (Fig. 7.13). Naola (1970) described the skeletal remains of the Samurai horses as being related to the Mongolian pony, that is, like the Chinese horses, they were stocky and had heavy heads.

8 Equids in the classical world

Fig. 8.1 Alexander the Great riding Bucephalus attacks Persian horsemen, on the Alexander sarcophagus, Syria. (Photo Hirmer Fotoarchiv).

Perhaps Bucephalus was the most famous horse ever to have lived. He is reputed to have been a black horse, bred in Thessaly and taken, in 343 BC, as a gift to Philip of Macedon by his friend, Demaratus. When led out for Philip to inspect, the horse bucked and reared and was unmanageable so that the king ordered him to be removed. Alexander, Philip's twelve-year old son, had seen, however, that the horse was shying at its own shadow; he took hold of the horse, turned his head so that he was facing into the sun and leapt on his back. Bucephalus was given to Alexander and was his favourite horse for the following seventeen years. No one but Alexander rode him and he was trained to kneel so that the king could mount him in full armour. Philip had predicted that Macedonia would never hold his son, and indeed Alexander the Great, although he died at the age of only thirty-two, conquered the lands from Greece to Afghanistan and founded the largest empire the world had yet known.

A number of legends explain why the horse was called Bucephalus, which means 'ox-head': perhaps he had a peculiarly wide head as noted by Strabo (15.1.29.); or perhaps he had the small bony projections that sometimes appear on the skulls of horses and are known as 'horns' (Lydekker 1912); or perhaps, as stated by Pliny

(VIII, 64), there was, 'the figure of a bull's head marked on its shoulder'.

After being with Alexander for seventeen years, Bucephalus died in 326 BC during the final battle against Porus, king of India, who fought with a huge cavalry of elephants. The horse was buried with great pomp and, in his memory, Alexander founded the now lost city of Bucephala, on a bank of the river Jhelum (today in Pakistan, Fig. 8.2) at the site of his last river crossing.

The exact age of Bucephalus was never known, but it was always claimed that he had been born in the same year as Alexander (Lane Fox 1973). The most famous monument to Alexander and his horse is the Alexander sarcophagus in Syria (Fig. 8.1). In a marble relief it shows the king, wearing the lion's head helmet of Heracles and riding Bucephalus in an attack on Persian horsemen.

Cavalry in ancient Greece

By the time of Alexander the Great, the battle-chariot, as used by the Assyrians for direct attack, had long since been supplanted, in the Greek states, in favour of the foot soldier and later the mounted archer. The use of large numbers of chariots lined up for shock-combat was a feature of war throughout the Bronze Age empires of Egypt, Crete, and the Near East. After the beginning of the historical period in Greece, around 700 BC, when iron technology had become widespread, and throughout the classical period until the end of the Roman empire, chariots were used mainly for transport to the scene

ig. 8.2 The travels of Alexander he Great who lived from 355–323 c.

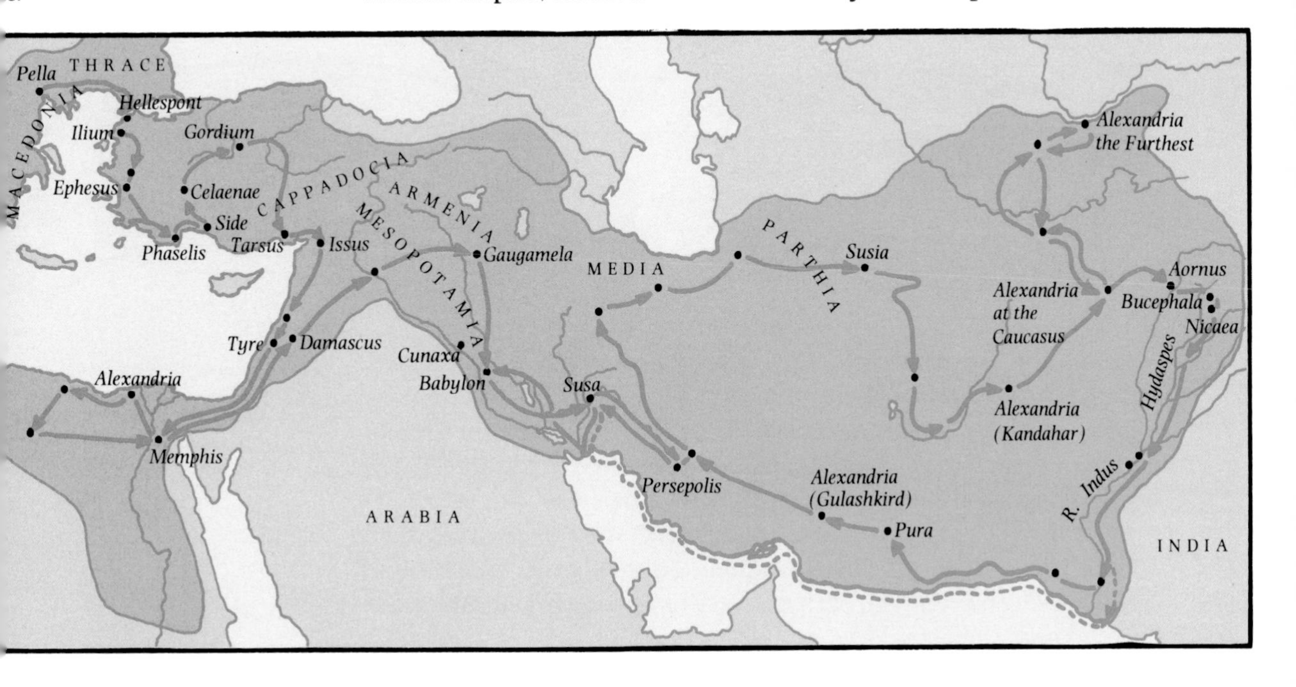

Fig. 8.3 Hoplites fighting on foot while their horses are held in the background, on a Middle Corinthian cup, sixth century BC. (Photo National Archaeological Museum, Athens.)

of a battle. Once the fighting began the chariot was left in charge of a driver and the warrior attacked on foot, with a javelin (Greenhalgh 1973).

Knowledge of much of life in ancient Greece, including methods of warfare, can be deduced from the magnificent paintings that have survived on classical vases. Greenhalgh (1973), among other scholars, has made a study of early Greek warfare based on an analysis of this art. It is known that until the latter half of the sixth century BC, nearly all battles were fought on foot by heavily armed soldiers who were known as hoplites (Fig. 8.3). As with chariot-riders, mounted hoplites would dismount at the front line, leaving their horses with attendants, and fight on foot (Fig. 8.4).

Fig. 8.4 A battle with cavalry and hoplites on an Attic vase. (Louvre, Paris, Photo Musées Nationaux.)

According to the laws of ancient Athens, every free man between the ages of eighteen and forty years had to be ready for military service whenever he was called on. The infantry made up the greater part of the army and consisted of three classes. The hoplites were the heavy troops and were armed with a pike or spear, dagger, body armour in the form of a corselet, and a shield. They were all free men and they formed the phalanx or main line of the battle, which could be up to sixteen deep with each man wielding a *sarissa* or pike made of wood from the cornel tree. In later periods these pikes were up to

21 feet (6.4 m) long and were a formidable defence against attack. Second were the light troops who were armed with javelins, but little defensive armour and no shields. They were mostly slaves who were used for skirmishing and supporting the hoplites. Third, were the mercenaries who carried javelins, bows and arrows, and slings for harrassing the enemy.

The cavalry was made up from the wealthiest members of society; it too was divided into heavy and light, the former carrying horse armour as well as body armour. Paintings on the Attic vases (Fig. 8.4) show that from around 550 BC the head-on attack by a cavalry charge against a phalanx of hoplites was beginning to be effective. The Greek horse was, however, only pony-sized and it could not carry armour heavy enough to repel spears hurled in large numbers by an infantry. Heavy armour is also impossible for a horseman riding without stirrups, which were unknown in the classical world (see p. 75), as it would impair his ability to grip, especially while throwing a lance. Greenhalgh (1973) pointed out, however, that although stirrups would give the lancer a more secure seat, they were not essential and horsemen often carried both lances and javelins.

The most famous writer on horses of the classical period in war and peace was Xenophon, a rich Athenian who lived from 428/7–354 BC, dying at the age of about seventy-four, a year after the birth of Alexander. Xenophon led the march of the ten thousand Greeks, in 401 BC, back to the Hellespont after the Battle of Cunaxa in Babylonia (Fig. 8.2) in which their former leader, Cyrus II was killed. Xenophon later wrote an account of the army's exploits under the title *Anabasis* which has been transcribed as *The Persian Expedition*, in the translation by Warner (1975).

It is clear from Xenophon's account that, although armed horsemen were well established at the end of the fifth century BC, they were not as yet a very important element in warfare. In one speech to the troops, Xenophon declared that:

> No one has ever died in battle through being bitten or kicked by a horse; it is men who do whatever gets done in battle. And then we are on a much more solid foundation than cavalrymen, who are up in the air on horseback, and afraid not only of us but of falling off their horses . . . There is only one way in which cavalry have an advantage over us and that is that it is safer for them to run away than it is for us.

There was undoubtedly a good deal of dissemblance in this speech as Xenophon himself belonged to the élite class of knights and was the owner of many horses. Later he attempted to organize a cavalry amongst his Greek followers in order to suppress attacks from Persian horsemen who were adept at the Parthian shot. That is, they would shoot backwards as they retreated on horseback (Fig. 5.12). Xenophon spoke thus:

> I have noticed, too, that we have horses in the army: some are mine, others are part of Clearchus's property which he has left, and there are many more which we have captured and now use for carrying baggage. If, then, we sort them out, putting baggage animals in the place of some, and equipping horses for the use of cavalrymen, they too, perhaps, will give the enemy trouble when he runs away. *Warner 1975 p. 160*

Even better known, perhaps, than Xenophon's account of the Persian expedition are his treatises on horsemanship and hunting (Marchant 1968) which, like those of the later Roman writers on husbandry, can only be described as models of good sense and compassionate care for the welfare of their valuable animals. Xenophon told the horseman everything he needed to know about how to choose a horse, how to mount, train, exercise, and feed it. He also gave advice on the type and size of armour to select for both horse and rider, and how to hurl a javelin, all, it must be remembered, without stirrups and with only a cloth for a saddle. Xenophon provides no evidence to indicate that stallions were castrated, which is probably why he advocates the use of a muzzle whenever a horse is led out of its stable without a bit. This would be because a stallion is much more difficult to control than a gelding and a group of stallions will readily bite and fight with each other. As a side-line Xenophon stated that the hair, ie the mane, tail, and forelock of mares was always cut off before they were mated by asses for the production of mules.

Apart from a few idiosyncratic comments such as this, every person who owns a horse today would benefit from reading Xenophon, as, no doubt, did Alexander the Great, one of the most consummate generals of all time. There is no surviving contemporary record of the life of Alexander but there are numerous later accounts, of which the most notable is by Arrian who lived from *c.* AD 95–175. Arrian, in his detailed description of Alexander's campaigns, often quoted from the lost writings of Ptolemy I who lived from *c.* 368–284 BC; he was a friend of Alexander's and became the first king of Egypt in 304 BC.

It is often difficult to assess the numbers of troops and horses involved in ancient battles, partly because the written accounts tend to be much exaggerated. However, Arrian, using the numbers given by Ptolemy, was probably in the right order of magnitude when he wrote that in the early spring of 334 BC Alexander marched to the Hellespont (Fig. 8.2) with an infantry that included not much above thirty thousand light troops and archers, and a cavalry of over five thousand. Later, in the famous Battle of Issus, against the last Persian king, Darius, the enemy troops were said to number 600 000 fighting men. Arrian gives a graphic description of the flight of Darius after this battle and it is noteworthy that he was still using a chariot:

> As for Darius, the moment his left wing was panic-stricken by Alexander and he saw it thus cut off from the rest of his army, he fled just as he was in his chariot, in the van of the fugitives. So long as he found level ground in his flight, he was safe in his chariot; but when he came to gullies and other difficult patches, he left his chariot there, threw away his shield and mantle, left even his bow in the chariot, and fled on horseback.
>
> *II. 11. 5–6. Brunt 1976*

Later, in 326 BC, in his final battle against the Indian king Porus at the river Hydaspes (Fig. 8.2), Alexander's five thousand horsemen (according to Arrian, V.14.6. quoting from Ptolemy) were matched against two thousand cavalry and a hundred and twenty chariots led by the son of Porus. In this first stage of the battle four hundred of the Indian horsemen fell, including Porus' son, and the chariots and their teams were captured, as they proved too heavy in retreat, and were useless in the action because of the mud. Porus then took 'all his cavalry, about four thousand horse, all the three hundred chariots, two hundred of the elephants, and any serviceable infantry, about thirty thousand, [and] advanced against Alexander'. All these troops were ranged in battle-order with the elephants in front. A desperate battle was fought and much carnage ensued of elephants, horses, and men, as described with vigour by Arrian. The result was:

> The Indians lost nearly twenty thousand foot, and up to three thousand horsemen; all the chariots were broken to pieces; two sons of Porus perished, with Spitaces, the nomarch of the Indians of this district, the commanders of the elephants and the cavalry and the generals of Porus' army to a man . . . and all the surviving elephants were captured. Alexander's army lost about eighty foot-soldiers at most out of a force which had been six thousand strong in the first attack; as for the cavalry, ten of the mounted archers, who were the first to engage, and about twenty of the Companions' cavalry with two hundred other troopers were killed.
>
> *V.18. 2–3. Brunt 1976*

It was at this battle that Alexander's horse Bucephalus died.

Arrian wrote about the Indians:

> Their horses are not saddled nor do they use bits like the Greek or Celtic, but a band of stitched rawhide is fitted round the muzzle of the horse, with bronze or iron goads, not very sharp, turned inwards. The rich use goads of ivory . . . Most of them ride on camels, horses and asses, the rich on elephants. For the elephant in India is a royal mount; second to it in dignity is a four horse chariot, and camels come third; to ride on a single horse is low.
>
> *16.11–12; 17.2. Brunt 1976*

By the time of Alexander, horses were being bred in huge numbers at special centres. Arrian wrote that there were originally about a hundred and fifty thousand mares kept on the Nesaean plain in Media (northern Iran), but that most of them had been driven off by robbers, leaving only fifty thousand. Horses on the Nesaean plain are also mentioned by Herodotus (VII. 40), writing around 450 BC and Strabo (64–63 BC–AD 25) wrote of the land near the new city of Pella in Syria on the river Orontes:

Here, too, Seleucus Nicator [formerly one of Alexander's captains] kept the five hundred elephants and the greater part of the army, as did later kings ... Here, too, were the war-office and the royal stud. The royal stud consisted of more than thirty thousand mares and three hundred stallions. Here, too, were colt-breakers and instructors in heavy-armed warfare, and all instructors who were paid to teach the art of war.

16. 2. 10. Jones 1983

Horses in ancient Greek pageantry and sport

Although the chariot was on the wane as a weapon of war in the classical world during the fifth century it certainly played a supreme role in all processions and in spectacles in the hippodromes. These were specially-built courses, notably at Olympia but also at other places where the sacred games were held, in which chariot and horse races played an important part (Fig. 8.5). The two-wheeled chariots were normally drawn by two or by four horses, in which case the two outside were held with traces. They were a symbol of wealth and it is easy to picture Greek dignitaries demonstrating their high status by riding about in their expensive chariots, at home and abroad, as for example in the description by Strabo of driving along the walls of the hanging gardens of Babylon:

Babylon, too, lies in a plain; and the circuit of its wall is three hundred and eighty-five stadia. The thickness of its wall is thirty-two feet; the height thereof between the towers is sixty cubits; and the passage on top of the wall is such that four-horse chariots can easily pass one another; and it is on this account that this and the hanging garden are called one of the Seven Wonders of the World. *16. 1. 5. Jones 1983*

Much can be learned about the religious processions from the sculptures of the Parthenon at Athens. The Parthenon is a Doric temple built of marble from Mount Pentelicus, between 447 and 432 BC. The frieze of stone reliefs that ran around the top of the outer wall represented the procession of the Great Panathenaea. This was the oldest and most important festival in Athens, and was originally a religious celebration. It was held, about the middle of August, in the

g. 8.5 Front view of a four-horse ariot on a Greek vase, *c.* 540 BC. hoto reproduced by courtesy of the ustees of the British Museum.)

third year of each Olympiad (see p 167), with a smaller festival in all other years. Many contests took place, including a chariot race of the *apobatai* who were the companions of the charioteers. They had to show off their skills by leaping out of their chariots while the horses were going at full speed. All this is depicted in the frieze, as described by Haynes (1975), and a number of horsemen and their chariots are shown in Figures 8.7–8.8. From the sculptures it can be seen that the horses were all rather small, probably being not more than 130 cm at the shoulder.

Cavalry in ancient Rome

Order, discipline, and hierarchy were the key words for the success of the great empire, which had its origins in a small urbanized area around Rome in the early third century BC. From around 290 BC to AD 117 the Romans succeeded in conquering all of western Europe,

Fig. 8.6 A centaur is about to throw a lapith by grappling his right leg, from South Metope XXXI of the Parthenon marble friezes, 447–432 BC. (Photo reproduced by courtesy of the Trustees of the British Museum.)

Fig. 8.7 A poorly preserved chariot with galloping horse from the South Frieze XXX 73–4 of the Parthenon marbles. (Pho reproduced by courtesy of the Trustees of the British Museum.)

Fig. 8.8 Horsemen leading the Panathenaic procession, from the West Frieze II of the Parthenon marbles. (Photo reproduced by courtesy of the Trustees of the British Museum.)

North Africa, and the Near East, as shown in Figure 8.9. From the first their armies were rigidly organized and highly disciplined.

The origin of the Roman legion is ascribed to Romulus who is claimed to have organized the Latin clans into three tribes, each of which had to raise 1000 foot soldiers and 100 cavalry. Later the whole population was numbered and divided according to individual wealth into six classes each with fixed positions in the army. The men had to pay taxes in the form of a war-tribute and had to pay for their own arms and armour. Every man between the ages of seventeen and forty-five was liable for service, but only the most wealthy went into the cavalry and were required to provide horses as well as arms and armour. Each legion normally had a total strength of 4500 men, of which 300 were cavalry.

Rather little is known about the breeding and training of horses for war, but Varro (II. vii. 14) made the comment that the experienced soldier would train his horse in one way, the charioteer and circus-rider in another, while the horse that was used as a pack animal needed to be docile and was usually castrated.

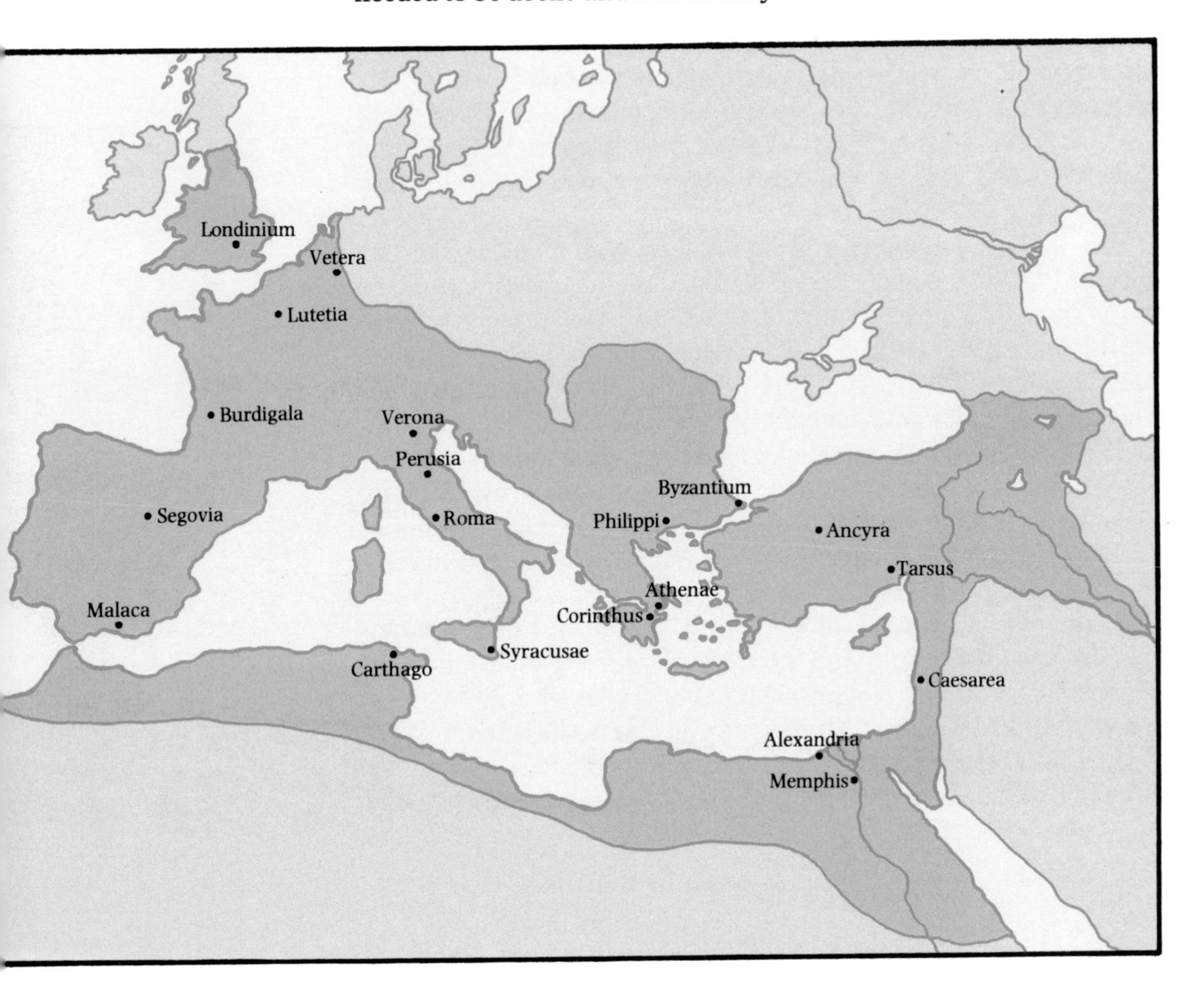

g. 8.9 The Roman empire in *c.* AD 7.

Julius Caesar (100–44 BC) was the only Roman general to leave a personal record of his campaigns, written by himself. His seven books (an eighth was added after his death) are written with great style and probably, considerable accuracy. During the Gallic wars, from 58–52 BC, Caesar had between six and eleven legions in Gaul and at least 4000 cavalry, all of whom were foreigners, either requisitioned from friendly or conquered tribes, or were mercenaries from Germany. The horsemen did not usually take an active part in fighting but were used for reconnoitres, for sending messages, and as a back-up for the infantry. Caesar was clearly in awe of the Germans and their horses, for in describing the Suebi tribe he declared (Handford 1951):

> Even horses, which the Gauls are inordinately fond of and purchase at big prices, are not imported by the Germans. They are content with their homebred horses, which although undersized and ugly, are rendered capable of very hard work by daily exercise. In cavalry battles they often dismount and fight on foot, training the horses to stand perfectly still, so that they can quickly get back to them in case of need.

In 55 and 54 BC Julius Caesar made two invasions of Britain, with a large fleet of ships carrying legions of infantry and cavalry. He crossed by the narrowest sea passage and landed near Dover. Of all Caesar's writing his accounts of these invasions are probably the most lucid, and they are well known to every schoolchild studying Latin.

The Gauls and the Germans did not fight with chariots, but this was the Britons' usual method of warfare and the invading Romans found them difficult to deal with. The British tactics were to drive wildly about, hurling their javelins which, with the galloping horses and the noise of the heavy wheels, threw the enemy into disorder. Then the warriors would jump down from their chariots and fight on foot. Meanwhile the drivers would retire and place the chariots in a position that was easily accessible for the warriors to rush back to. Even this unfamiliar method of warfare, used by very large numbers of charioteers on their own ground, did not prove invincible against the might and discipline of Caesar's army.

By the time of the final conquest of Britain in AD 43 the Romans had become all-powerful and the network of paved roads that had been built across the empire meant that troops and government officials could be moved around speedily and efficiently (Sitwell 1981). The state hierarchy, organization of the army, and the Roman roads are all part of the heritage of the western world today.

Equids in Roman agriculture and transport

It is evident from the Roman writers on agriculture, Columella and Varro, that the ox was the most commonly used beast of burden and plough animal. Mules and asses were the next most popular animals for traction while horses were used mainly for riding, warfare, and for racing in the circuses (see p. 168).

Columella (VI. xxvii.1) divided horses into three stocks, the noble which supplied horses for the circus and the Sacred Games; the stock for breeding mules, which he claimed, in the price that its offspring fetched, was as valuable as the noble stock, and the common breed which produced ordinary mares and horses. The utmost care was taken with feeding of the first two; they were only allowed to mate in alternate years and at the time of the spring equinox so that the foals would be born a year later when the pasture was at its most lush. But common horses were pastured everywhere together and there was no season fixed for their breeding. The Romans were able to assess the age of a horse with considerable accuracy from the stages of eruption and wear of the teeth.

Male donkey foals that were considered suitable for breeding mules were removed from their mothers and fostered onto a mare. Thus the donkey became more familiar with the behavioural patterns of the horse than the ass and could be more readily persuaded to mate with a mare when it grew old enough to breed.

Both Columella and Varro described the interbreeding of domestic donkeys with wild asses. Columella (VIVI. xxxvi.4) claimed that the 'mules' of Africa regularly produced fertile offspring, which was presumably because they were not hybrids between a donkey and a mare, but between a domestic donkey and a wild ass (*Equus africanus*) which is the same species. Varro (II. vi. 3–3) wrote that the many herds of wild asses in Phrygia and Lycaonia (Anatolia), called onager (*Equus hemionus*), were well suited for breeding with domestic donkeys.

Varro recommended that the young donkey begin work when it was three years old and he explained that:

> ... there really are no herds of these animals except of those which form pack trains, for the reason that they are usually separated and sent to the mills, or to the fields for hauling, or even for ploughing where the land is porous, as it is in Campania. The trains are usually formed by the traders, as, for instance, those who pack oil or wine and grain or other products from the region of Brundisium or Apulia to the sea in donkey panniers.
>
> *II.vi.5 Hooper & Ash 1967*

The domestic donkey was spread all over the Roman Empire as a pack animal and was probably even brought to the British Isles. The mosaic of Silenus riding on a realistic portrayal of an ass, shown in

Figure 8.10, suggests that the artist was very familiar with this animal, and a few bones from archaeological sites of the third century AD in southern England have been identified as donkey. Although the donkey was the favoured pack animal the hybrid vigour of the mule gave it a higher value for long-distance transport and for the haulage of carts (Fig. 8.11).

In efforts to control the shortages of food, fuel, and many other resources as the Roman Empire declined, ever more stringent laws were passed in Rome about almost every activity and way of life of the people. A compilation of these laws, which were decreed by the emperors from AD 313 to AD 438, in the reign of Theodosius II, has come to be known as the Theodosian Code. Despite its complicated political and social structure the huge Roman Empire depended entirely on oxen, horses, mules, and donkeys for all its land transport and postal services. The laws of the Theodosian Code show how necessary it was to protect the post horses and wooden carts from

Fig. 8.10 Silenus riding on an ass on a Roman mosaic found under Dyer Street, Cirencester, England in 1849. (Photo Corinium Museum, Cirencester.)

ig. 8.11 Coin showing a Roman ule car or *carpentum*. This was a vo-wheeled covered cart that was nly used by a state-priestess on feast ays, as well as very occasionally by woman in the imperial family. aius Caligula gave this honour to is dead mother, Agrippina the Elder, AD 37–41, when he instituted ames in her memory. Her image ould have been carried in rocession on the cart when the ames were held. (Photo reproduced y courtesy of the Trustees of the ritish Museum.)

overloading. All the passages quoted below are from Pharr (1952):

A weight of one thousand pounds shall be loaded upon a postcarriage and of not more than six hundred pounds on a cart, with the additional provision that gold and all other supplies of the largesses shall be conveyed, not on vehicles selected according to the desire of the official escorts and tax receivers, but on vehicles that are suitable for the burden and weight. Of course, under the threat of capital punishment, they shall not be permitted to place on these vehicles any private burden, otherwise than is prescribed by Our law . . . .

Since also the management of the posthorses must be treated in a similar way, the saddle and bridle shall not exceed sixty pounds in weight, and the saddlebags shall not exceed thirty-five pounds, with the provision that if any person should exceed the weight limits prescribed by the imperial moderation, his saddle shall be cut into bits and his saddlebags shall be assigned to the account of the fisc. *17 June*, AD 385 (8. 5. 47)

Eight mules shall be yoked to a carriage, in the summer season, of course, but ten in winter. We judge that three mules are sufficient for a two-wheeled conveyance. *24 June*, AD 357 (8. 5. 8)

The weights appear to be rather low, considering that the Roman pound was equivalent only to 0.72 of an English pound or 0.34 of a kilogram, but the Roman horse was still only pony-sized and unshod, while the weight that the wooden axle of the cart could bear was strictly limited.

By the middle of the fourth century AD, disintegration of the Empire was beginning, imperial bureaucracy held the whole population in a vice-like grip, from which escape to bands of brigands and cattle-thieves could be a strong temptation. For this reason pastoralists were not allowed to own horses:

In order that all efforts of brigands may cease because of lack of resources, We deny the right to possess a herd of horses to the shepherds of the estates of Our privy purse, that is, to the herders of wool-bearing sheep and of cattle. *5 October*, AD 364. (9. 30. 2)

In additon, in an attempt to preserve the dignity of the city of Rome and to prevent infiltration by brigands, everyone was to wear the toga or other clothing suitable to their station, and 'no person shall be allowed to appropriate to himself the use of boots or trousers'. Also 'All dignitaries of high civil or military rank shall have the right always to use within Our City of most sacred name the vehicles of their rank, that is, two-horse carriages.' (14.12.1).

Although punishments for offenders were extremely severe the last emperors, including Theodosius II who was a Christian, appear to have been disturbed by the cruelty of gladiatorial shows and laws were passed to stop them:

Bloody spectacles displease Us amid public peace and domestic tranquility. Wherefore, since We wholly forbid the existence of gladiators, You shall cause those persons who, perchance, on account of some crime, customarily sustained that condition and sentence, to serve rather in the mines, so that they will assume the penalty for their crimes without shedding their blood. *15. 12. 1*

Compassion for animals is a feature that runs all through the writings of the ancient Greeks and Romans and it is not surprising since, not only did the wealth of the people rest in their livestock, but animals and people lived so closely together. Cattle, sheep, and equids would have been stabled in one end of the family home and every aspect of their daily lives would be familiar to their owners. However, it is revealing that an emperor had thought for retired chariot horses, in this law:

We decree that provender from the fiscal storehouses shall be furnished to the Palmatian and Hermogenian horses, when they have been weakened by their lot as contestants in the chariot race or by their number of years or by some other cause . . . *1 January*, AD 371 (15. 10. 1)

Obtaining enough horses for the Roman cavalry and enough mules and donkeys to transport the legionaries' arms and baggage round their Empire must have been a continual problem to those in power, despite such measures as demanding a stock of military horses to be brought from North Africa as part of the regular taxes. Lack of sufficient horsepower is likely to have been one deciding factor in the final overthrow of the Romans, around AD 460, by mounted Franks, Goths, and Vandals whose bowmen were far more experienced at fighting on horseback than the Romans.

9 Equids in the Middle Ages

The horse in feudal Europe

ig. 9.1 A horseman of the eventh to eighth century AD in tone relief, probably part of the hancel of a church at Hornhausen, ermany, discovered in 1871. Photo Landesmuseum für orgeschichte, Halle.)

After the fall of Rome to the Vandals in AD 410, there was a period of turmoil throughout Europe; not only was there a general collapse of the Roman Empire, but population movements and changes in tribal structure led to the break up of the small kingdoms of Celts, Germans and Slavs, and the beginnings of organization into national states. This was the period of the Dark Ages, the Anglo-Saxon and Viking explorations, and the supremacy of the Gothic and eastern cavalries (Fig. 9.1). It was also the time of the legendary King Arthur and the beginnings of feudalism and chivalry*. Innumerable feats of mounted warriors are recorded in the many epic poems and sagas, of

*The term chivalry is used for the edieval knights or men-at-arms ho were equipped for battle and ommitted to a code of religious, oral, and social practices.

which the little known *Gododdin*, written around AD 600, is one of the earliest. It is an account of a disastrous cavalry raid into Yorkshire from Edinburgh, by Mynyddog, chief of the Gododdin tribe, as described by the poet Aneirin, who was one of the few survivors:

> Bleiddig son of Eli was a wild boar for fierceness; he drank off wine from brimming glass vessels; and on the day of combat he would do feats of arms, riding his white steed. Before he died he left behind him bloodstained corpses.
> *Jackson 1969*

The feudal system, which dominated the politics of the Middle Ages in Europe, was based on the relationships between landowners and their tenants. It was a way of organizing society for instant warfare by large numbers of feudal lords, or knights, who replaced public authority, and had complete power over their tenants (serfs or vassals). Throughout Europe, from Britain to Hungary, political power was held in the grip of these military lords, but each country had its own manner of waging war, inherited from the mosaic of tribal peoples who had overthrown the Roman Empire. In the west, with the Franks, that is the Germans and French, and the Anglo-Saxons, mounted warriors were still in the minority. In the east, the Vandals, as well as the Ostrogoths, Visigoths and Lombards (Fig. 9.2) had been accustomed to fighting on horseback for hundreds of years, and their strategies for invasion were to continue into the tenth century.

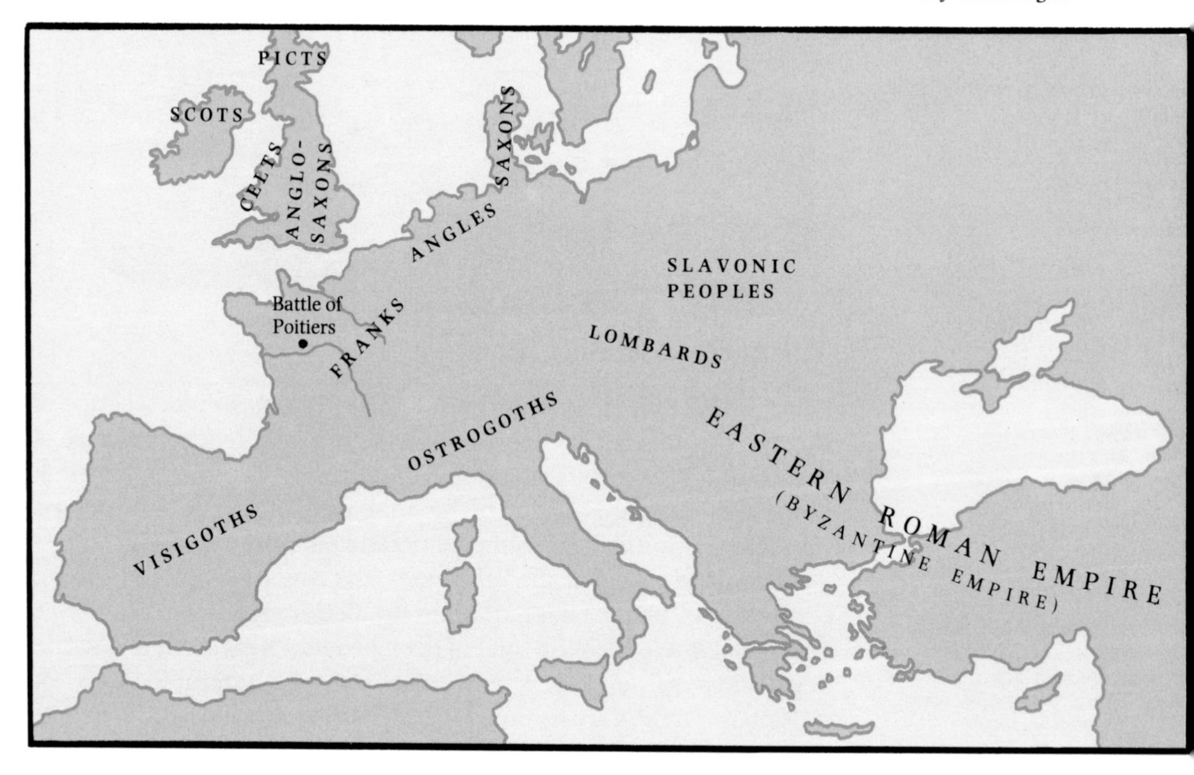

Fig. 9.2 The peoples of Europe in the early Middle Ages.

ig. 9.3 Bronze statuette of a knight n horseback (an *aquamanile* for vashing the hands at table), from Iexham, Northumberland, England, hirteenth century. (Photo eproduced by courtesy of the 'rustees of the British Museum.)

The Battle of Poitiers (see also Fig. 9.2), fought on 25 October 732, is generally taken as the turning point in history for the feudal armies of the west. After Charles Martel had routed the Muslim cavalry at this battle, the composition of the Frankish armies underwent a drastic change and the horseman took over from the foot soldier as the most important military unit. Whether this was in great part due to the increased power and efficiency afforded by riding with the innovations of a saddle and stirrups has been discussed at length by White (1962) and Contamine (1986).

From this time on, the single role of the élite class of knights was to be armed horsemen, a way of life which was extremely expensive. Armour was becoming increasingly complicated as the art of war became more and more elaborate, so that special horses had to be bred or imported to carry the weight of the knight and his weapons (Fig. 9.3). In the eighth century military equipment for one man cost as much as twenty oxen, that is the plough teams of ten families, according to White (1962).

The size of the Medieval war-horse has long been a subject of contention. Legend holds that the knights' horses were mighty steeds and implies that they were as large as modern cart horses, but no remains of such animals have ever been found, earlier in time than the fourteenth century. The remains of horses from archaeological sites of all periods before the late Middle Ages indicate that they stood, at the most, 14 hands (142.2 cm) at the shoulder. Their rather small stature would not, however, prevent the horses, which were always uncastrated stallions, from carrying a heavy load, especially as the tactics of fighting were by shock combat at close range so, once at the line of battle, the cavalry did not have to travel over great distances.

The great raids of the Magyars

Eastern Europe, during the tenth century, was in the grip of raids from the Magyars who held a military kingdom of several united tribes in Hungary. They were armed horsemen who lived by raiding and their tactics were the same as those of the Mongol hordes. Their archers made a sudden attack, accompanied by the terrifying war cry of '*Hui hui*'. Then they pretended to withdraw, only to wheel around on their pursuers. The Magyars lived in waggons and tents and made forays right across Europe into France. They were only subdued when Henry of Saxony came to the throne of Germany in 919. He became known as Henry the Builder because he entirely reorganized the people of his country into fortified towns and trained them for mounted warfare.

Henry defeated a Magyar invasion at the Battle of Allstedt on the 15 March 933 and in 955 Henry's son, Otto the Great, with eight thousand horsemen, finally drove them away at the Siege of Augsberg (Oman 1924 1).

The art of venery

In its archaic meaning, *venery* was the practice of hunting beasts of game, often known as the chase. The word was derived from the Latin *venari* to hunt, and it came into common English usage in the early fourteenth century. Prowess in the art of hunting as practised by the feudal élite of Europe was an important aspect of social life, as well as an essential training for chivalry and war, and in times of peace it was a substitute for the battle (Thiebaux 1967; Fig. 9.4).

Although there is a relatively prolific Medieval literature on venery, and large numbers of paintings and tapestries of hunting scenes, there is remarkably little information on the horse in the chase. This is surprising as the knights were always mounted for the hunt, and there are detailed descriptions of the different breeds of dogs that were used for hunting with instructions for their welfare (see, for example, Baillie-Grohman 1904).

VERPAGE

ig. 9.4 Hunting and killing the volf, from *Le Livre de Chasse* by Gaston Phoebus, Count of Foix, who ived from 1331–1391. (Photo Bibliothèque Nationale, Paris.)

ig. 9.5 The 'shield-wall' of the Anglo-Saxons in the Battle of Hastings fought on 14 October 1066, rom the Bayeux tapestry, Bayeux Museum. (Photo Michael Holford.)

ig. 9.6 The Normans carried ten horses in each ship across the English Channel for the invasion of William he Conqueror, from the Bayeux apestry, Bayeux Museum. (Photo Michael Holford.)

ig. 12.6 The Byerley Turk, painting by John Wootton. (Photo E. T. Archive. See p. 172)

One rather sad mention is quoted by Cummins (1988) from the French royal hunting accounts of 1398. This being expenses claimed, 'For 4 carcasses of old, worn-out horses bought in Nemours market to feed several thin and ailing hounds kennelled at Fontainbleu'.

The Battle of Hastings and the Bayeux tapestry

The loss of England to the Normans on 14 October 1066, at the Battle of Hastings, is one of the landmarks of European history. Almost every incident that occurred, from the political background over the previous seven months and the earliest planning of the battle to its final moments, are as well known as those of any armed encounter in history. The source of this knowledge is in written accounts such as those of the *Anglo-Saxon Chronicles*, but perhaps most notably in the Bayeux tapestry. This is a roll of stitchwork, in several colours, that was probably made by English women shortly after the time of the battle.

The tapestry measures more than 200 feet (60 metres) in length and is 19.5 inches (49.5 cm) wide. According to Douglas & Greenaway (1953) it depicts 623 persons, 202 horses and mules, 55 dogs, 505 other animals, 37 buildings, 41 ships and boats, and 49 trees. These portrayals are extraordinarily vivid and show many details of ship-building, dress, armour, weapons, and harness.

Oman (1924) described the battle as 'the last great example of an endeavour to use the old infantry tactics of the Teutonic races against the now fully-developed cavalry of feudalism'. The people of England, as they seem to have been throughout history, 'were prepared for the last war', for they were still fighting with the military methods of the seventh century. Although both the Norman and Anglo-Saxon armies rode with stirrups, as can be seen in the tapestry, when it came to the battle, the Anglo-Saxons dismounted and fought on foot, forming a 'shield-wall', which in the old Germanic style was impervious to attack from an enemy infantry, but when assailed by mounted archers was quickly mown down (Fig. 9.5).

The troops, brought across the Channel by William, Duke of Normandy, numbered around 7000 men, half of whom had mounts (Contamine 1986). Their transport and provisions must have required massive organization and a very large number of ships. Ten horses could be carried in each ship, as shown in the tapestry (Fig. 9.6), which means there must have been 350 ships for the horses alone. Besides the fighting men and their horses there was also the train of servants and baggage, carried by mules and pack-horses, which had to be ferried over the Channel.

9.5

9.4

6

12.6

According to tradition, King Harold heard of the landing of the Normans, on the first or second of October, while he was at York celebrating his victory against Hardrada and the Norsemen at the Battle of Stamford Bridge (fought on 25 September). He immediately rallied his army and by a forced march managed to reach London on the seventh or eighth of October. Two further days' march brought him close to the enemy camps. Oman (1924), using contemporary accounts, has described the battle in great detail: how the English host was ranged on the crest of Telham hill above Senlac marshes, how it was armed with all manner of primitive weapons, and how the Norman duke and his knights, drawing on their heavy mail-shirts at the last minute, formed three parallel corps, each containing infantry and cavalry (Fig. 9.7). 'Desperate as was their plight, the English still held out till evening; though William himself led charge after charge against them and had three horses killed beneath him, they could not be scattered while their king still survived and their standards still stood upright.'

In the end, Harold received a mortal wound from an arrow in his eye; the king fell to the ground, and the remnants of the English army fled in disarray.

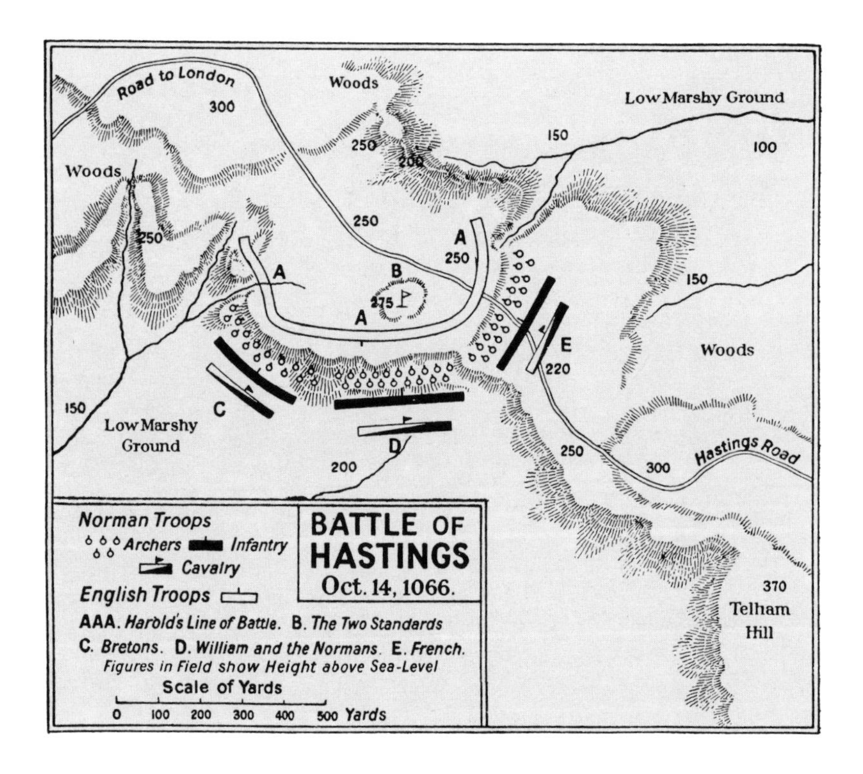

Fig. 9.7 The Battle of Hastings, from Oman (1924).

ig. 9.8 A charge with couched ances. From the *Book of St Edmund.* Twelfth century. (Photo Pierpoint Morgan Library, Ms 736 fo. 7v.)

The Crusades

By the end of the eleventh century the supremacy of the armed horseman was firmly established throughout Europe. All lands, except those of the extreme north, were relatively settled and the knights began to look further afield for extension of their power (Fig. 9.8). As early as the first Crusade, 1096–1099, the motive was the conquest of territory either by the ruling knights, such as Bohemond, leader of the Normans, who wanted lands in the east, or by the pious pilgrims who marched on foot because they wished to see the walled city of Jerusalem returned to Christian rule. Then there were the Italian republics to whom the coastal plain of Syria and its ports were of the utmost importance for commerce.

Like all wars, the Crusades were fought over possession of territory, while religious fervour and the principles of chivalry ensured that there was always sufficient manpower and horsepower to supply the huge armies needed for the innumerable battles and skirmishes with the Turks, Arabs, and Egyptians. According to historians from Oman (1924) to Contamine (1986) the journeys of the Crusaders were the most badly organized and mismanaged military expeditions of all time. This was in part due to arrogance and lack of preparation by the many leaders who had no central unity, but also to lack of any geographical knowledge, despite the network of Roman roads that was still in use.

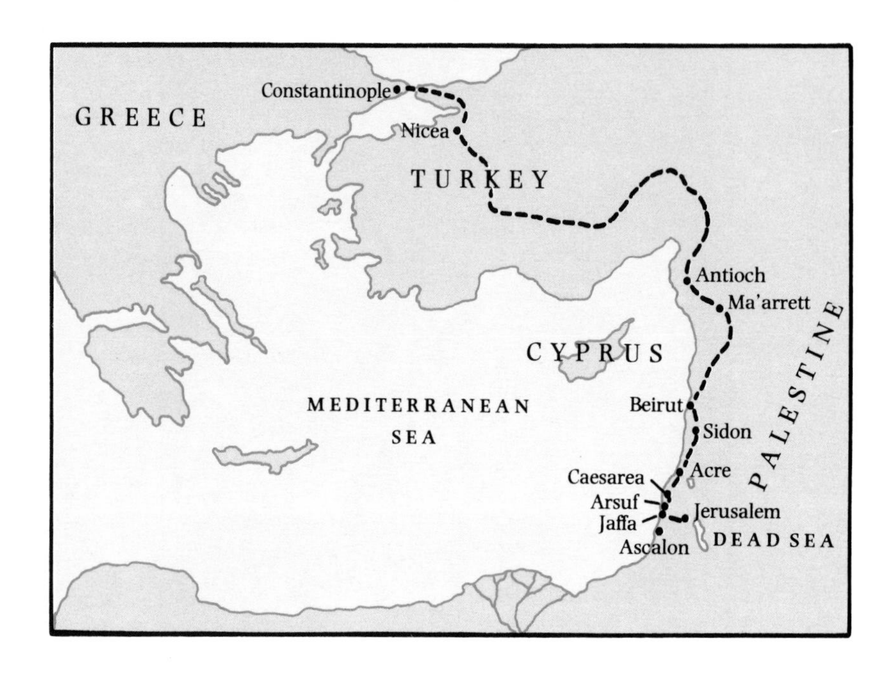

Fig. 9.9 The region of Palestine with Crusader sites and the route of the first Crusade, 1096–9.

The first Crusade was much the most successful in terms of military victories. In May 1097 entry to Turkey was achieved through the Battle at Nicea (Fig. 9.9), and the following autumn the Christians began their invasion into Syria. Until the following June, Antioch (now Antakya in Turkey) was beseiged and won despite heavy odds against the Crusaders. Their next aim was the Holy City of Jerusalem which fell on 15 July 1099 and which they held until 1187.

The final victory of the first Crusade was the Battle of Ascalon (now in Israel) in August 1099. It was against an army sent from the Fatimid kingdom of Egypt to defend the cities of the Syrian coastline which had been under its dominion. The Egyptians were defeated because, like the Anglo-Saxons at the Battle of Hastings, they fought on foot. Small (1987) has discussed the different manner of fighting by the various Muslim armies involved in the Crusades. The Turks never remained stationary but pursued a strategy of encirclement of the enemy, even when they were outnumbered. By this means their expert horsemen assailed the enemy from all sides. The Arabs also fought on horseback, but according to Small they did not have the

mobility of the Turks, and their equipment was heavier and more like that of the Christian Franks. The Egyptian army or 'Aethiopes' was made up of Arabs, Berbers, and Sudanese horsemen, with bowmen on foot who provided a solid target for the charges of the mailed and mounted Frankish knights.

One hundred years later, on 7 September 1191, Richard Coeur de Lion, with his Crusaders, fought and won the famous battle of Arsuf (now in Israel) against the Turks led by Saladin (Fig. 9.10). This battle followed the typical strategies of the period, which depended greatly on the troops being able to hold a solid front. A line of footsoldiers stood firm in front of the mounted troops who were divided into squadrons of around a hundred horsemen. The battle itself began at a fairly slow pace and it was not unusual for the King

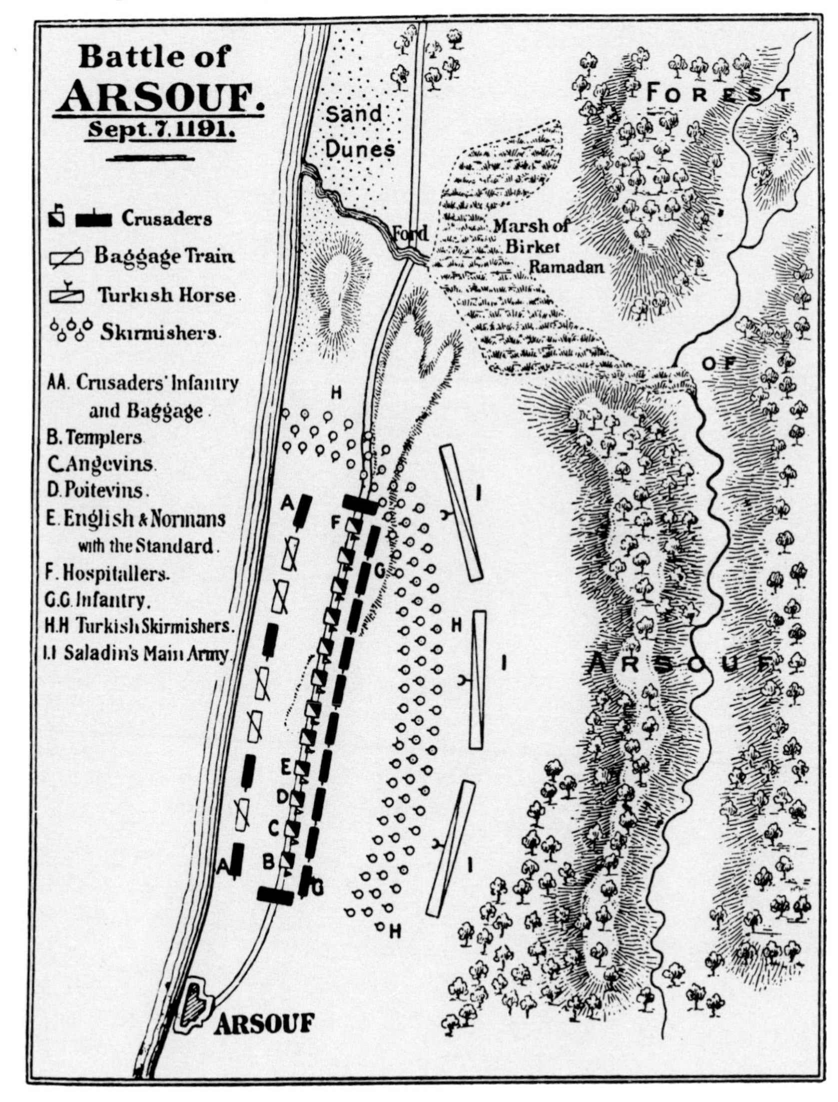

ig. 9.10 The Battle of Arsuf, from man (1924). Note Oman's spelling f Arsuf

and his closest knights to be followed by grooms with spare horses. The Royal Standard of England was fixed on a covered waggon drawn by four horses in the centre division of the army.

The following graphic account of the start of the battle of Arsuf has been quoted from the *Itinerarium Regis Ricardi* by Oman (1924 1):

> All over the face of the land you could see the well-ordered bands of the Turks, myriads of parti-coloured banners, marshalled in troops and squadrons; of mailed men alone there appeared to be more than twenty thousand. With unswerving course, swifter than eagles, they swept down upon our line of march. The air was turned black by the dust that their hoofs cast up. Before the face of each emir went his musicians, making a horrid din with horns, trumpets, drums, cymbals, and all manner of brazen instruments, while the troops behind pressed on with howls and cries of war. For the Infidels think that the louder the noise, the bolder grows the spirit of the warrior.

Winning the battle of Arsuf gave the Franks the whole coast-land of southern Palestine, but they were continually harassed by the horsemen of the Mamelukes and Kurds, led by Saladin, and they never achieved their goal of re-gaining Jerusalem.

Fig. 9.11 The empire of Genghis Khan, *c.* 1227

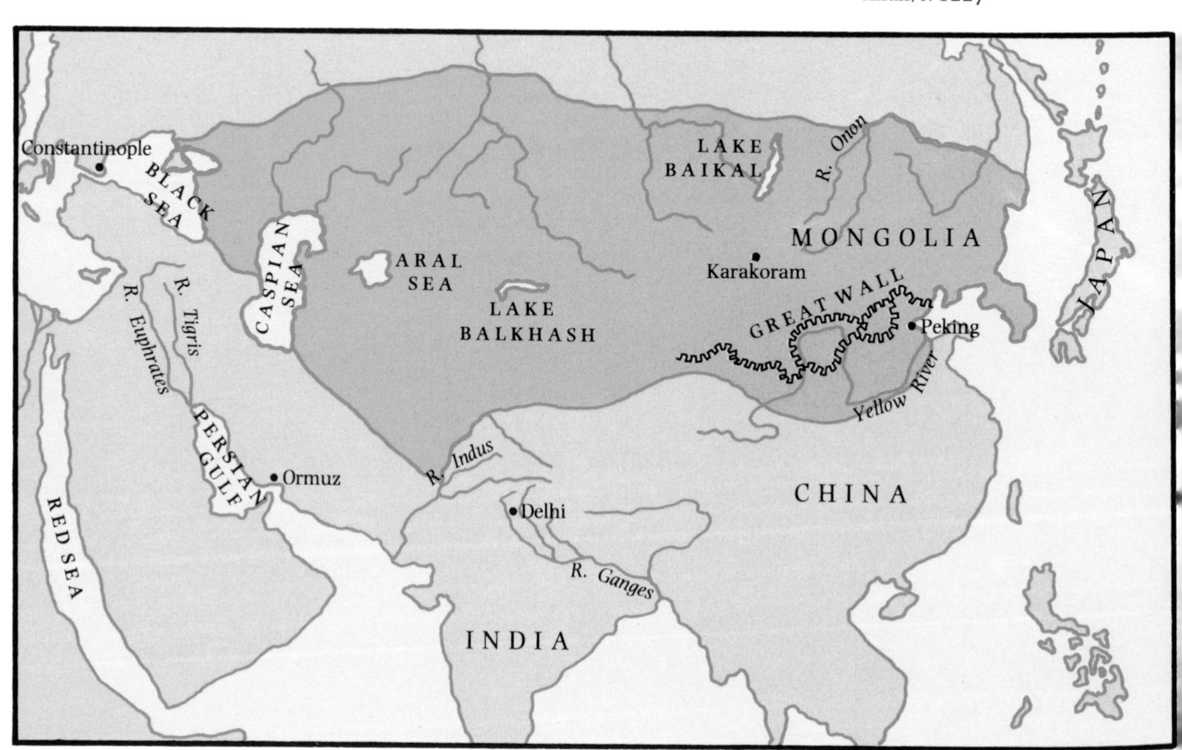

ig. 9.12 It has been claimed that his saddle and stirrups with laborately carved ivory decorations elonged to Genghis Khan.

Genghis Khan (1162–1227) and the Mongol hordes

Thirty years before the battle of Arsuf, Genghis Khan, the most powerful nomad and one of the greatest conquerors the world has ever seen, was born in a tent on the banks of the river Onon in Mongolia (Fig. 9.11). He was named Temujin after a Tartar chieftain who was defeated in battle by the baby's father, the Mongol emperor Yesukai, just before the birth in 1162. Temujin was thirteen years old when his father died. His mother, Yulun, was determined not to let go of the throne, so Temujin was surrounded by unceasing warfare amongst the rival nomadic tribes, until in 1206, at the age of forty-four, he proclaimed himself ruler of the empire and adopted the name of Genghis Khan which, in Chinese, means 'ruler of all'. He conquered northern China in 1215 and then advanced westwards, always on horseback and with appropriate pomp and ceremony as evidenced by the rather uncomfortable-looking saddle with carved ivory pommel that he is supposed to have owned (Fig. 9.12).

The world of Genghis Khan became a nomad empire which extended from the Mediterranean to the Pacific (Fig. 9.11). It was based on the tribal loyalties of pastoralists and nomadic horsemen who, being reared in the saddle, had little experience of urban civilizations, but were unified and manipulated by the remarkable organizing ability of their single leader, the Khan.

The socio-economics of the Mongolian nomads and their relationships with the sedentary societies of central Asia in the Middle Ages have been discussed by Khazanov (1983). That these nomadic herdsmen, through their warrior élite, could subjugate settled states over such a huge area, was indeed a unique event in history. The aristocratic warriors have been compared to the feudal lords of

Medieval Europe, but Khazanov (1983) contends that there was little similarity in their methods of taxation and social control.

The story of Genghis Khan and his four sons is one of unrivalled military success, due in great part to the supreme training of their horsemen. The Mongol hordes quickly earned a reputation for the ruthless slaughter of all their enemies which was probably a major contribution towards their seeming invincibility. One of the few instances of mercy shown by Genghis Khan was to the Turk Jelaleddin who made a daring escape from a battle on the banks of the river Indus. When all was lost Jelaleddin seized a fresh horse and jumped into the river, twenty feet below, and was watched by the admiring gaze of Genghis Khan as he emerged on the opposite bank, dripping wet but unharmed.

The descendants of Genghis Khan were less ruthless and his son Ogödey, according to Khazanov (1983), understood the aphorism of the ancient Chinese orator Lu Tsia (third to second century BC) who declaimed: 'Although you inherited the Chinese Empire on horseback, you cannot rule it from that position'.

The travels of Marco Polo (1254–1324)

To the general reader of history, the peoples of the Middle Ages appear to have been endlessly travelling great distances; there were the Vikings, then the Crusaders, then the Mongol hordes, and then there was Marco Polo, who, without an army, travelled the furthest, across Eurasia to China, and as far south as Java (Latham 1988).

Marco Polo was the son of a Venetian merchant who had already made one journey to China when he set out again, in 1271, accompanied by his brother and Marco, aged seventeen years. They spent the next twenty years travelling around the empire of Genghis Khan's grandson, the great Kublai Khan (1260–94) (Fig. 9.13).

Marco Polo's account of his travels is extraordinarily detailed, and it is possible to gain a great deal of information not only about the way of life of the numerous different peoples he lived with, but also about the domestic animals and the entire natural world around him. Whether the whole of *The Travels* is a true account of Marco Polo's experiences has been questioned, but there can be no doubt that his observations on natural history were accurate. The way of life of the people he met had remained little changed since ancient times, and the nomads of Mongolia, in particular, appear to have retained many of the customs of the ancient Scythians. These included, apparently, the burial of the later Khans in the Altai mountains, accompanied by the ritual killing of humans and horses.

While in Persia, Marco Polo saw fine horses that were exported to India where apparently they could not be bred because of the heat.

He also reported on the asses, 'The finest in the world, which eat little, carry heavy loads, and travel long distances, enduring toil beyond the power of horses or mules' (Latham 1988).

In China, Marco Polo described how the Great Khan left his palace every year on the 28th day of August to inspect his stud of snow-white stallions and more than 10 000 snow-white mares. With the exception of the Horiat tribe, no one who was not of royal lineage was allowed to drink the milk of these mares. The white horses were treated as sacred animals and when they were grazing, 'if a great lord were going that way he could not pass through their midst, but would either wait till they had passed or go on until he had passed them' (Latham 1988).

The method of travel of the nomad armies was highly organized and the horseman carried no baggage, except:

> They each carry two leather flasks to hold the milk they drink and a small pot for cooking meat. They also carry a small tent to shelter them from the rain. In case of need they will ride a good ten days' journey without provisions and without making a fire, living only on the blood of their horses; for every rider pierces a vein of his horse and drinks the blood. They also have their dried milk which is solid like paste.

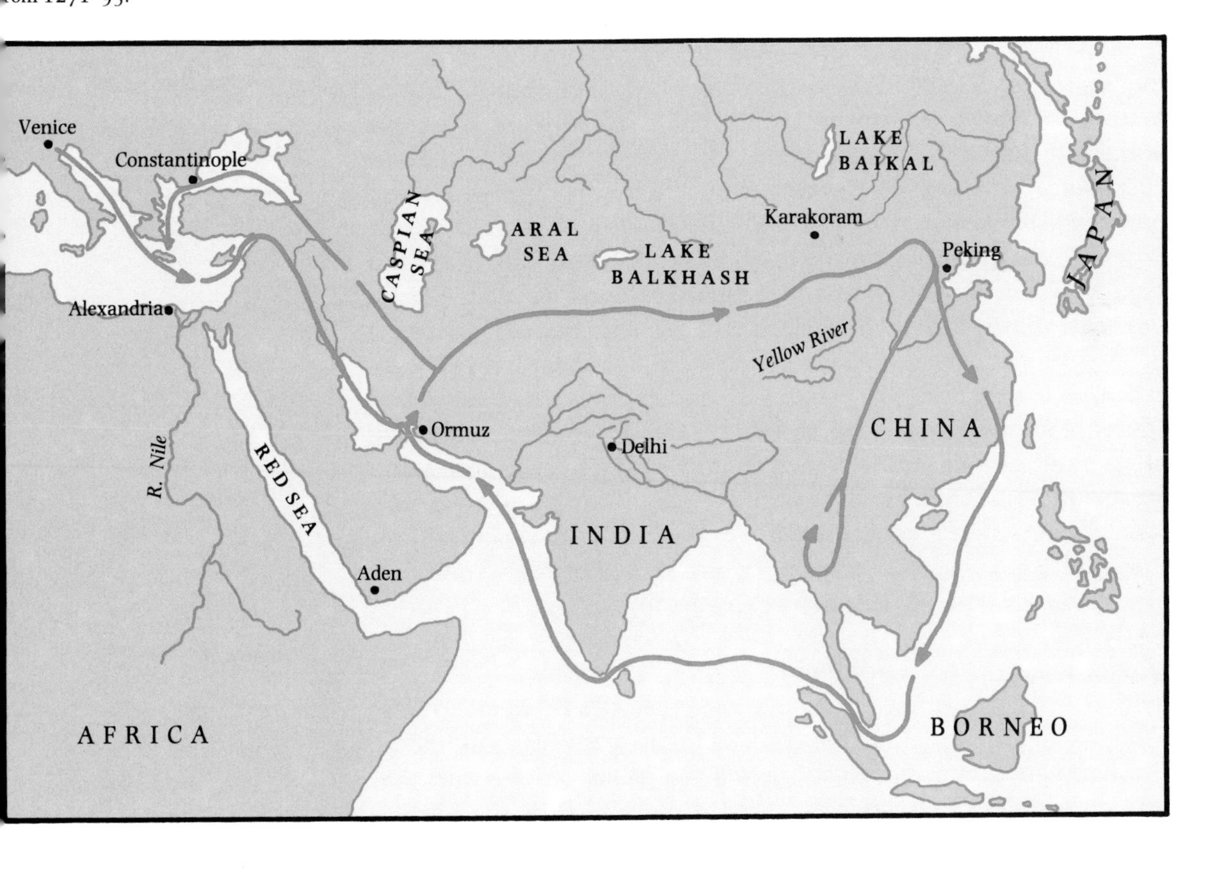

'ig. 9.13 The travels of Marco Polo rom 1271–95.

Perhaps the most remarkable of all Marco Polo's anecdotes are his descriptions of the postal and messenger services that were set up to keep the Great Khan in touch with the whole of his vast empire and to bring him exotic fruits from distant parts. There were relays of human runners who could cover two hundred miles in a day, by each man running his fastest for three miles, and there were relays of mounted messengers who covered twenty-five miles at a stretch.

The posts were stationed twenty-five or thirty miles apart along all the main highways, and there were claimed to be more than 10 000 of them. Each post was equipped with a luxurious rest house and a complement of four hundred horses, which were provided by the inhabitants as part of the system of taxation. Two hundred of the horses were put out to graze on the pastures while the other two hundred were held ready to be taken on to the next post by the messengers. Marco Polo's comment was that, 'the whole organization is so stupendous and so costly that it baffles speech and writing'.

The economics of horsepower in the Middle Ages

IN THE EAST

Marco Polo's explanation of how the Great Khan's postal services could be financed was that the population was enormous, which he maintained was partly due to the custom of each man having as many as ten wives or more, so that he could have up to thirty sons, and partly due to the great areas of cultivated lands that yielded an abundance of rice and millet. Presumably there was still enough land per head of human population to allow such affluence, and to provide enough grazing for the enormous herds of horses that were required for battle as well as for transport. Marco Polo claimed that the beasts of the Tartars multiplied without end and that when on military service each man had eight or more horses for his own use.

But, if it was ever so, it was a short-term period of glory and affluence for, as described by Khazanov (1983), the traditional way of life of the Mongolian nomads went into sharp decline during the fourteenth century, and wives and children had to be sold into slavery. Across the great empire of the descendants of Genghis Khan there were continual internecine wars, struggles over land, and discord between the settled peoples and the nomads.

IN THE WEST

In western Europe, during the Middle Ages, there were no vast expanses of grazing lands that could support great herds of horses. Warfare was extremely expensive and the provision of horses and armour could only be funded by the taxing of serfs by their feudal lords. The first priority for the farm horse was probably to provide a breeding stock, the finest animals being allocated to the knights for battle and for hunting, the ponies being used for transport and haulage.

The Welsh Laws of Hywel Dda, written around AD 940, provide valuable information on the status of the horse and other animals at this early period (quoted below from Probert 1823, but see also Owen 1841):

> The value of a young foal of a mare, is four pence until fourteen days after its birth . . . If it attain the age of one year and a day it is worth forty eight pence. If it attain three years of age, it is worth sixty pence; and then it will be time to tame it with a bridle and to teach it its duty, whether as a stallion, a palfrey, or a serving horse . . . the value of a palfrey [riding horse] is one hundred and twenty pence. The value of a pack horse, is one hundred and twenty pence. The value of a serving horse is sixty pence. The value of a stallion is the price of his two stones [testes], with two mares, and he himself forming the third; that is, one hundred and eighty pence . . .
> The value of a filly is four pence for the first fourteen days after her birth; sixteen pence when a year old; thirty two pence when two years old; and then she must be set to work . . . her qualities are to draw a car up steep ground and down it, and to breed colts, and if she do not possess these qualities, one third of her price must be returned.

> Neither horses, mares, nor cows ought to be yoked to the plough; and if they should be so, and the mares and cattle suffer an abortion from it, there can be no reddress.

> Four horse shoes and their complement of nails are worth two pence.

> The price of a saddle is eight pence; a bridle gilt with gold, is eight pence; gilt with silver, six pence; and stained, darkened, and done with copper, four pence. The price of a pair of spurs, gilt with gold, is four pence; gilt with silver, two pence; and darkened, stained, and done with copper, one penny . . . if the stirrups be gilt with gold, their price is eight pence; if with silver, four pence; and if done with copper, or darkened or stained, four pence. The surcingle [girth] is of the same value as one of the stirrups; the two girths are two pence; the horse cloth, four pence; a saddle cushion of plaited work, one penny; one of linen, one penny; and a covering is one penny.

There are two most interesting points to be noted from these laws. First, horseshoes and stirrups were, by this time, clearly taken for granted as part of the equestrian equipment, and second the use of horses for ploughing was still in disfavour and there is only the one mention of its occurrence, quoted above, while in other sections of the laws there are numerous references to ploughing and plough oxen.

Although horseshoes, saddles, and stirrups were in use for riding horses in the tenth century there is still no mention of the horse collar which was to make the horse a more efficient draught animal, as discussed in Chapter 5. However, the new harness of a collar over the withers, a breast strap, and lateral traces, was in general use over the whole of northern Europe by the twelfth century (Fig. 9.14).

Fig. 9.14 Three horses, harnessed with horse-collars, pulling a cart, from the Luttrell Psalter, *c.* 1320–1340. Remarkably, the driver of the cart does not appear to be a man but a monkey or ape wearing a hat (probably the tailless monkey, known as the Barbary ape, *Macaca sylvanus*). (Photo British Library.)

White (1962) put forward the interesting view that, between the eleventh and thirteenth centuries, many peasant farmers, who had previously lived in small hamlets close to their fields, began to congregate, further from their land, in villages of two or three hundred families. He suggested that this was made possible by the new use of horses for ploughing, because they could move more quickly than oxen between the farmers' houses and their fields.

In the early Middle Ages, estimates of how many men and horses could be called on for war at any one time are very contradictory and the true numbers probably varied greatly and would be impossible to assess. Contamine (1986) has suggested that in England in 1298 with an army of 25 700 infantry and at least 3000 knights, perhaps five per cent of the adult male population had been called to arms. Likewise, it is impossible to give a general estimate of the numbers of casualties after a pitched battle but, again quoting Contamine (1986), average figures of around fifty per cent of the men and horses seem not unreasonable.

It was, of course, not only the knights that had to have horses, but every army of importance needed hundreds of baggage animals, and, when there was a shortage of horses, mules, donkeys, and oxen were used. Contamine (1986) has discussed the high value of horses, the capture of war horses, and the trade in importing horses from Spain and Italy to France, despite the more general custom of riding mules in these countries.

It is true to say that, just as the Mongol hordes depended on their horses for survival, so did the strengths of the western nations depend on the numbers of their mounted horsemen in war, and the numbers of transport and baggage animals that could be raised to support them. In some countries, during the fourteenth to mid-fifteenth centuries the cavalry took on even greater importance and increased in numbers relative to the infantry. For example Contamine (1986) quotes figures of 8000 cavalry and 3000 infantry to be maintained, during war time, by Venice in 1485. This was the heyday of the armoured horseman who was soon to give way to the growth of artillery with the new weapon of gunpowder.

10 Conquest of the Americas

Crosby (1972), quoting Vazquez de Espinosa, relates that at the beginning of the seventeenth century on the grasslands of Tucumàn, in Argentina, there were wild horses 'in such numbers that they cover the face of the earth and when they cross the road it is necessary for travellers to wait and let them pass, for a whole day or more, so as not to let them carry off tame stock with them.' These were horses that were living and breeding in the wild, after their introduction to South America by the Spanish, little more than a hundred years previously.

This poses one of the most intriguing questions of historical biology: why were these horses so successful when their antecedents had become extinct on similar grasslands at least 8000 years earlier?

In 1965 a conference was held in Boulder, Colorado, to discuss the phenomenon of the mass extinctions of animal species throughout the world at the end of the last ice age, around 10 000 years ago. The proceedings of this conference were published as *Pleistocene Extinctions: the Search for a Cause* (Martin & Wright 1967). This book has become a classic that has profoundly influenced the interpretation of human activity in the past. Its radical thesis was that the primary cause of the extinctions was over-hunting by humans, which destabilized the populations of large mammals, already under stress from drastic climatic and environmental changes.

Over the past twenty years the theory of over-hunting as the main cause of Pleistocene extinctions has become more generally accepted. Yet there are still difficulties, especially with the large mammals of North and South America, in seeing how human predators could have exterminated such a great diversity of species in a relatively short period and over such a vast area. The reasons for the extinction of the several species of equid at this period are a particular enigma when considered against the success and rapid expansion of horses, living in the wild, on both these continents, after their introduction as domestic animals by the Spanish in the fifteenth century. Yet the facts are incontrovertible: all the species of endemic wild equid were extinct in the western hemisphere by 8000 years ago, and by AD 1700 all the grasslands and prairies, south of the Arctic, were inhabited by enormous numbers of feral domestic horses.

The extinct species

Finds of fossil horses have been described from the Americas since the middle of the last century and they have been given innumerable names, a new name often being based on a single tooth. It is not surprising that there should be a large number of species since equids first evolved in North America and they underwent an extensive evolutionary radiation. However, during the Pleistocene, two main lines evolved from the diversity of horse-like animals of the Miocene and Pliocene. These were the three-toed horse, *Hipparion*, which first appeared in North America but, before becoming extinct there, moved through Europe to Africa where it survived well into the Pleistocene. The second line was *Dinohippus*, the ancestor of all living equids which also migrated into Asia, there evolving into the true horse, *Equus ferus* (see p. 17: Fig. 1.1).

Fossils representing all the living species of equids except the true horse (*Equus ferus*) have been identified from North America, these being the zebra, ass, and onager (Dalquest 1978). However, during the late Pleistocene all these became extinct, along with many other species of herbivores and their predators. In North America these included the mastodon, mammoth, camels, yak, saiga antelope, sabre-toothed cat, and dire wolf. In South America, the giant sloth is perhaps the most notorious extinct mammal and its remains have been found from a number of sites associated with human activity (Martin 1967).

From where and when humans first travelled to North and South America is still the subject of controversy. But, if the radiocarbon dating is correct of two out of eight hearths at the site of Pedra Furada in north-eastern Brazil, it was probably at some time around 32 000 BP (Guidon & Delibrias 1986). Further evidence for human occupation at this very early period may be forthcoming from the waterlogged site of Monte Verde in the rainforests of south-central Chile which was excavated by T.D. Dillehay between 1976 and 1985 (publication forthcoming). Most other radiocarbon dates for the earliest archaeological sites in the Americas centre around 18 000 BP and this is coincident with the beginnings of the megafaunal extinctions, which by 8000 years ago had resulted in the loss of thirty-five genera (McDonald 1984).

At the end of the last glaciation (the Wisconsin) in the western hemisphere forests greatly increased over both continents, as in Europe. At the same time, low-level forage would also have been increased with the warming of the climate, and it is hard to see why this Holocene environment should have been detrimental to grazing mammals, and the equids in particular. Again as in Europe (see p. 11), it can be postulated that human hunters disrupted populations

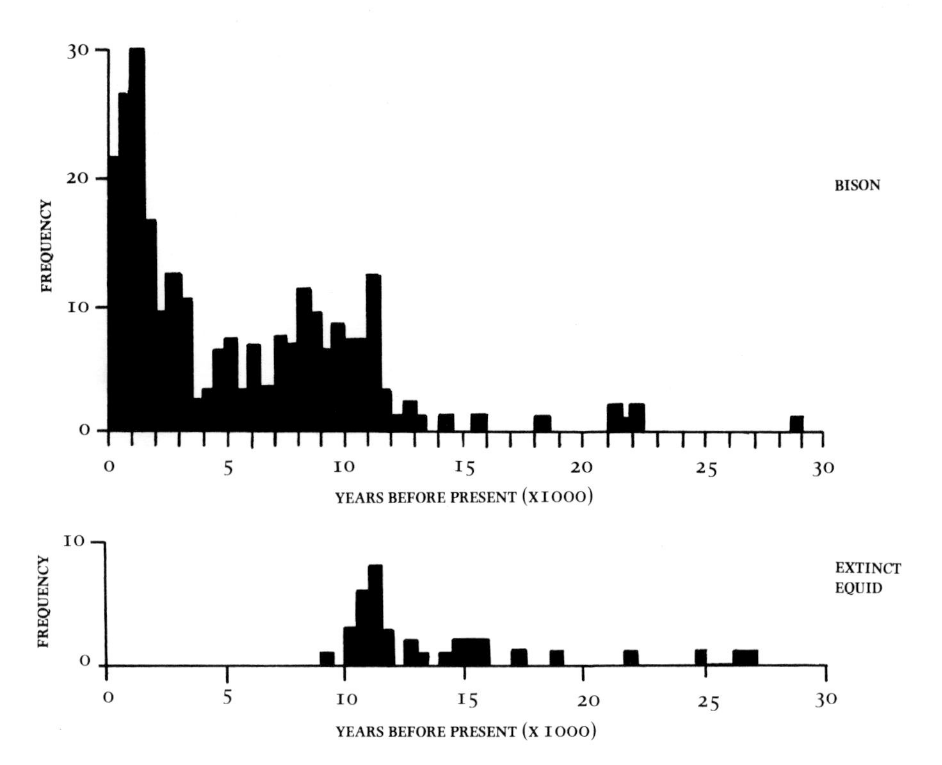

Fig. 10.1 Plots to show the numbers of radiocarbon dates that have been obtained for finds of biso and extinct equids from sites in Nor America. After McDonald (1984).

of equids and other slow-breeding mammals, notably the mastodon and mammoth, which were already in decline as a result of environmental changes and competition from other proliferating species such as the bison and deer.

The remains of equids are very common from archaeological sites before 10 000 BP in North America but there is no evidence that they were slaughtered in huge numbers at one time as were bison in the drives by native Americans after the end of the last ice age. The replacement of equids (and other extinct herbivores) by the bison as a provider of meat and other resources for human hunters is reflected in the pattern of radiocarbon dates obtained on their remains, as shown in Figure 10.1.

The extinct equids of South America have received less attention from palaeontologists than those in North America but three to four species have been described from Lower Pleistocene sites in Argentina. These equids apparently had short, heavy limbs with very large heads and were given the name *Hippidion* Owen, 1869 (Fig. 10.2). From the late Pleistocene, bones of a similar equid, named *Onohippidium* Moreno, 1891, were described from Eberhart Cave, Ultima Esperanza, where they were found in association with human remains (Martin & Guilday 1967).

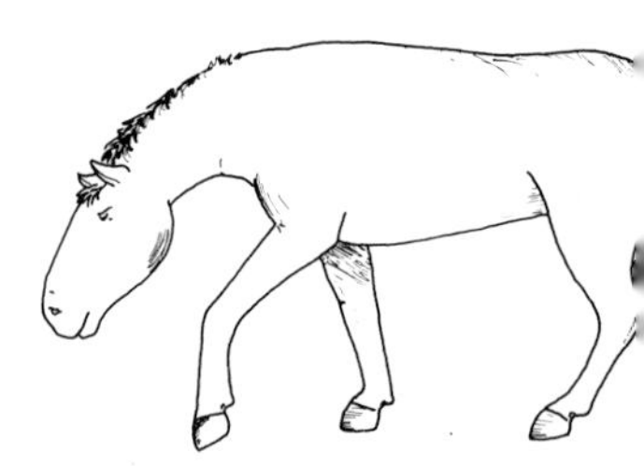

Fig. 10.2 *Hippidion*, the extinct equid from South America (see Figure 1.1). Drawing based on find of the skeleton, after Martin & Guilday (1967).

Introduction of the domestic horse

After a gap of 9000 years the grasslands of the western hemisphere were again to be grazed by herds of wild horses. But these were of foreign origin and they had a long and complicated history of continental migrations, matching that of the conquistadores who had brought them to the New World. For these herds were descended, initially, as are all domestic horses, from wild populations that were first tamed in Eurasia around 4000 BC. After five thousand years of breeding as domestic animals they were to return to the wild in the continents of their early evolution (see Fig. 1.1). Three vast regions of the Americas provided grasslands suitable for expansion of the feral horses: the prairies which stretch north from Mexico right up to Canada, the llanos of northern South America (Venezuela and Colombia) and the pampas, south of Brazil in Argentina and Uruguay.

The first voyage to the western hemisphere by Christopher Columbus set sail from Spain in 1492, with horses on board (Fig. 10.3). The ships of all later voyages were loaded with horses but so many died on the sea crossing that the part of the ocean between

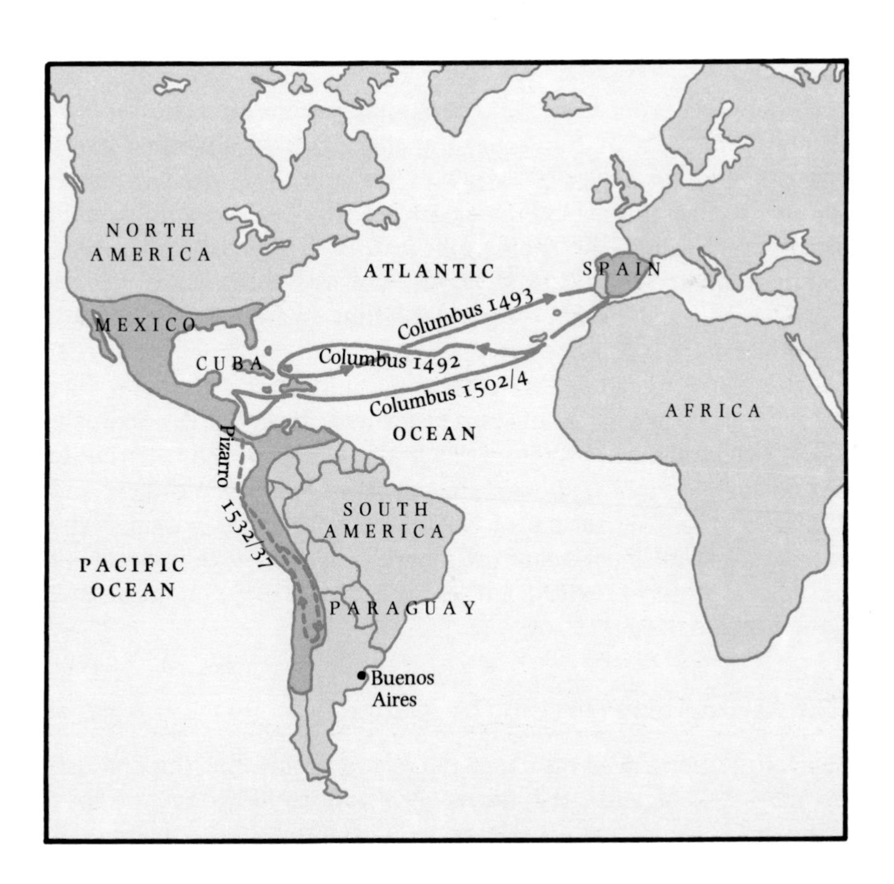

Fig. 10.3 Spanish invasions of the Americas.

Spain and the Canary Islands was called the *Gulfo de Yeguas*, the Gulf of Mares. At the same time the Atlantic tropics became known as the Horse Latitudes because so many horses died and were thrown overboard in this windless belt on the edge of the north-east trade winds. However, by 1503 there were sixty to seventy horses on the island of Española (Hispaniola), according to Crosby (1972).

The first region of the North American continent to be colonized by the Spanish was around Mexico City, known as New Spain, where there was good grazing for livestock. Although horses were slow to breed at first, within a few years of 1550 there were ten thousand horses in the area of Querétaro and they spread north from there with unimpeded progress (Crosby 1972).

In South America, Pizarro travelled south to Peru on horseback in 1532, evoking terror and amazement in the native Americans who had never seen such large animals, which 'ran like the wind and killed people with their feet and mouths'. At the same time, the first settlers arrived on the east side of the Andes, and it was not long before the Spanish and their horses reached the Argentine grasslands.

The city of Buenos Aires was first founded by Pedro de Mendoza in 1535 but he was forced to abandon the settlement through lack of food. He and his compatriots are claimed by Hunter (1837) 'to have passed over to Paraguay by water, leaving behind five mares and seven horses'. From these, and presumably from more horses that escaped into the wild from other travellers, there must have been a great population explosion. For it is reputed that when the next group of settlers arrived in Buenos Aires, led by Don Juan de Garay in 1580, they found the plains (in further words of Vazquez de Espinosa, quoted from Crosby 1972): 'covered with escaped mares and horses in such numbers that when they go anywhere they look like woods from a distance'.

Crosby (1972) claims that when the English arrived in America they called the plains 'deserts', and left them alone, but the Spanish, who already had a tradition of cattle ranching, turned their cattle out on the grasslands, allowed them to breed in vast numbers and controlled them by means of the horse. The cattle provided the expanding Spanish population with as much meat as they could eat but above all they provided leather for industry and enough tallow for lighting in the gold mines.

Native Americans become horsemen

Fear of the horse was the most effective weapon that the Spanish invaders had against the native civilizations (Fig. 10.4). They encouraged the native Americans to believe that horses were gods which had to be bridled to keep them from devouring humans, and there is a story that the horse of Hernàn Cortes was deified by the Mayans. Cortes, with a small army, had landed on the coast of the Mexican Gulf on 12 March 1519 and over the next two years he totally destroyed the Aztec empire. During this time his horse, named *El Morzillo*, became lame and he had to leave it with the Mayans. It soon died, but when European missionaries visited the place about a hundred years later they found a temple, in the centre of which was a great statue of a horse. This, of course, the missionaries duly destroyed (Barclay 1980).

Slowly the native Americans of both continents began to realize the value of the horse and they learned its management from the Spanish. Probably the first horses were obtained by stealing, and horsemanship was learned as much by trial and error as by copying from the Europeans. By the beginning of the seventeenth century the mounted Indian with his armoury of arrows and lance was more than a match for the Spanish rancher with his single firearm, and there were enough horses for every Indian family to own several.

g. 10.4 The conquest of Mexico, ɔm the sixteenth century *Lienzo de ɑxcala*. (Photo reproduced by ·urtesy of the Trustees of the British ·useum.)

Nomadic horsemen in North America

Before they obtained horses, the only transport the peoples of North America had, apart from the backs of the women, was the canoe and the dog sleigh or travois, and they hunted buffalo (the North

American bison, *Bison bison*) by driving them on foot. However, once they could ride horses the native Americans became masters of the buffalo and their hunting techniques and warfare assumed new patterns and rituals not unlike those of the Medieval knights of Europe. These have been described in the detailed work by Ewers (1955) on the role of the horse in the culture of the Blackfoot Indians of the northern plains of Canada. Horsemanship had spread to the Blackfoot by 1740 and soon became highly sophisticated. Each hunter had a specially-trained buffalo horse which did not fear the buffalo and was extra intelligent and surefooted. The horse had to ride close up to the buffalo and as soon as it heard the twang of the bow-string it swerved away, so that it was well out of harm's way when the wounded animal turned on its attacker (Barclay 1980; Fig. 10.5).

Hunting of the buffalo was done either by a surround or in open chase. In the surround, a large number of horsemen encirled a herd and milled around it, shooting down animals as they rode amongst them, while the chase involved a straight rush by mounted men who each singled out an animal to shoot and then rode alongside it for the kill (Ewers 1955). A skilled hunter mounted on a buffalo horse could kill enough animals in a morning to feed a family group of twenty as

Fig. 10.5 The Buffalo Hunt by George Catlin, 1841. (Photo reproduced by courtesy of the Trustees of the British Museum.)

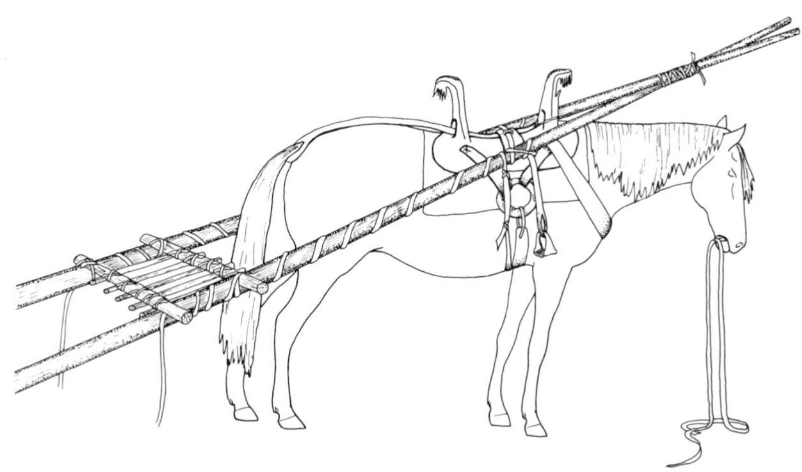

ig. 10.6 A native American horse arnessed to lodge poles. After Ewers (1955).

well as their dogs and leave enough meat for drying. A successful hunter therefore had plenty of leisure time for caring for his horses, making weapons and raiding enemy camps. Buffalo hunts were controlled by very strict social rules. There were severe penalties for anyone who hunted buffalo before the appointed time. Amongst the Cheyenne there were only three recognized crimes; homicide, disobeying the rules of the buffalo hunt, and repeated horse theft; for any of these the culprit was severely beaten.

According to Ewers (1955) an average Blackfoot family (two adult males, three adult females and three children) would have owned at least twelve horses; one to carry the lodge cover and its equipment, two to drag the lodge-poles, two for carrying the meat and other food, three to carry the women and children, two common riding horses for the men, and two buffalo horses (Figs. 10,6, 10.7). In addition there were usually four or five spare horses. A man who owned forty horses was considered to be rich and one who owned less than five was poor.

ig. 10.7 The breaking in of a ative American horse. After Ewers (1955).

The Navaho owned around 10 000 horses and in total, the 120 000 or so people of the plains owned about 160 000 horses in 1874 (Barclay 1980). These numbers can be compared with those of horses owned by the nomads of Tuva in southern Siberia in the last century. Most of these nomads were very poor, but there were wealthy individuals who owned a thousand or more horses (Vainshtein 1980). However, it must be remembered that the native Americans had only horses and dogs, while the Mongolians were true nomadic pastoralists and owned herds of reindeer, camels, cattle, sheep, and goats, as well as their horses. A closer similarity can be seen between the native Americans and the reindeer herders of the Eurasian tundra, the difference being that the reindeer hunters used domesticated deer to hunt wild deer of the same species, while the Americans used a new and introduced domestic species to hunt an endemic wild bovid. In neither case would the people eat the meat of their domestic animals unless they were exceptionally hard-pressed for food (Ingold 1980).

Whether the adoption of horse-riding, of itself, would have led to a huge increase in the numbers of buffalo killed by the native Americans is arguable. Before the arrival of Europeans with their firearms the bison provided the people with an inexhaustible supply of meat, leather, and almost all the resources they required. The great bison-kills (preserved in the archaeological record by the remains of hundreds of skeletons piled on top of each other) may not have been very common events, but even so the ability to kill selected animals by singling them out from a herd would result in a lower overall mortality. Moreover, ownership of horses enabled family groups to move away more readily from areas where food was becoming scarce and, in this way, as long as there was abundant land this prevented over-hunting in any one area. So within the one hundred and forty years of the supremacy of the horse, the buffalo herds might have remained stable, if it had not been for the gunfire of the European hunters between the 1860s and 1880s, whereby untold millions of buffalo were slaughtered and their tons of bones ground up for fertilizer. By the end of the nineteenth century the buffalo was almost extinct, the structure of social life amongst the tribes of native Americans was destroyed, and their horses were valueless.

The horsemen of South America

As with the native Americans in the north, the Pampas and Patagonian peoples of the southern continent had perhaps less than two hundred years during which they could be described as having a culture based on the horse. In a manner strikingly similar to that of the Khoi San of southern Africa at a somewhat later period, they first obtained their horses and other livestock by raids on the immigrant Europeans, but there were soon so many horses that they became common property. In 1596, in Paraguay, it was decreed that the wild horses should belong to anyone who took the trouble to catch them.

Many native South American peoples who had been horticulturalists gave up this way of life and took to hunting animals such as rheas, guanaco, deer, and peccaries, as soon as horses became abundant after the beginning of the seventeenth century. With the ability to travel further, women could gather different plants for food and medicines and the whole structure of the South American cultures of the grasslands changed. However, as in North America, the change was short-lived.

The South American peoples, unlike those in the north, always ate a fair proportion of horse-meat and horse sacrifices became common. In a ritual reminiscent of those of the ancient Scythians (see p. 97), when a Patagonian chief died, four horses were killed, their skins were stuffed and they were propped up on sticks at the four corners of the grave. The flesh was then eaten. When an ordinary man died his horse was killed and set up in the same way with its head facing towards the person's grave (Barclay 1980).

A detailed account of the wild horses and the people of the Pampas was written in 1802 by the Spaniard, Don Felix de Azara, and selections of this work were published in English translation by Hunter (1837). He added extensive footnotes from the, 'rough (but true to life) Notes' of the Englishman Sir Francis Bond Head from a time before the Indians commonly owned firearms. He wrote:

> The occupation of their lives is war, which they consider as their noble and most natural employment . . . The principal weapon which they use is a spear eighteen foot long; they manage it with great dexterity, and are able to give it a tremulous motion, which has often shaken the sword from the hand of their European adversaries . . . .When they assemble, either to attack their enemies, or to invade the country of the Christians, with whom they are now at war, they collect large troops of horses and mares, and then, uttering the wild shriek of war, they start at a gallop. As soon as the horses they ride are tired they vault upon the bare backs of fresh ones, keeping their best until they positively see their enemies. The whole country affords pasture to their horses; and whenever they choose to stop, they have only to kill some mares. The ground is the bed on which from their infancy they have always slept, the flesh of mares is the food on which they have ever been accustomed to subsist, and they therefore meet their enemies with light hearts and full stomachs . . .
>
> How different this style of warfare is from the march of an army of our brave, but limping, foot-sore men, crawling in the rain through muddy

lanes, bending under their packs! while in their rear, the mules, and forage, and pack- saddles, and baggage, and waggons, and women – bullocks lying on the ground unable to proceed &c. &c., form a scene of despair and confusion, which must always attend the army that walks instead of rides, and that eats cows instead of horses. *Hunter 1837*

The native Americans of the pampas were indeed hardened horsemen who lived a nomadic existence and were renowned for their bravery. They rode bareback, sometimes by clinging on, almost under the horse's belly, and often without even a bridle. They slowly succumbed, however, to European diseases such as smallpox and measles, and to the debilitating effects of alcohol, while the horses were all but exterminated by European firearms because they were interfering with domestic stock by leading away the mares and taking the grazing from cattle. This slaughter is described by Azara (Hunter 1837) who believed that the wild horses were 'useless and prejudicial'.

Although Azara may have approved of the wholesale extermination of the wild horses he correctly observed their ecology and behaviour. He noted that:

Each stallion appropriates to himself as many mares as he can, and takes care of them, keeping them always united, and fighting with any of his brethren who strive to dispute his possession of them. Hence it results that each herd of wild horses is composed of a multitude of small troops, a little separated, or almost united *Hunter 1837*

Almost a hundred and fifty years later, Berger (1986) described the formation of bands in the feral horses in the Granite Range of northwestern Nevada in North America:

An obvious feature of horse societies is year-round bands – a grouping pattern found in all populations that have been studied for at least six months, as well as in two species of zebra . . .

Most studies have found that bands consist of females, their young, and a single stallion . . .

The most impressive concentration of wild horses I observed was 210 animals in 1977. They grazed peacefully in adjacent bands southeast of Hart Mountain, Oregon.

Azara also commented that all the wild horses he saw were either light or dark chestnut, or bay in colour (Hunter 1837). If a white, dark, or piebald horse was seen with a wild herd it was certain to have been a strayed domestic horse. Berger (1986), on the other hand, found that wild horses occur in many colours, 'duns, bays, pintos, palaminos, sorrels, chestnuts and more'.

In Azara's time the feral horses of the pampas were called *baguales*. Today, the small dun-coloured feral horses on the high plateaux of

Colombia are probably all that remain of the original descendants of those taken to the continent by the Spanish in the sixteenth century (Lever 1985). In North America, however, there are many more feral horses, known as mustangs, and there are populations numbering some thousands in many states, as well as in Canada. Those that are believed to be of nearly pure-bred Spanish origin are conserved by a number of breed societies (Sponenberg 1984).

Herds of feral donkeys or *burros* are also fairly common in many western states where they compete with the desert bighorn sheep (*Ovis canadensis*) for forage, and therefore it is necessary to cull them (see also Chapter 2; Lever 1985).

The ethos of the cowboy and the ranch

Ranching is the use of large areas of land for the grazing of livestock, usually cattle or sheep, which are then sold. The livestock live as wild animals that may be rounded up only twice in their lives, first so that they can be branded and second for slaughter. During the last century, on the grasslands of North America, feral cattle, notably the Texas longhorn, exploded in numbers to fill the ecological niche left vacant by the extermination of the buffalo.

The earliest European ranchers on the North American plains had what seemed to be boundless land and an infinite number of cattle to be exploited for their hides and tallow. The meat was at first of little value because it could not be preserved. The territorial behaviour of the feral cattle meant that each herd maintained a home range where pasture, shade, and water were available, so the rancher only had to isolate himself with a herd or herds and prevent his stock from intermingling with others. This was done by rounding up the cattle on horseback, branding them, and then keeping them together.

This phase of loose control was short-lived, for with the coming of the railways the grasslands were connected to urban markets and there was a large influx of new cattlemen who were not prepared to share the pasture, which was coming close to being over-grazed. What had been ownership of cattle by appropriation became ownership of land, and the invention of barbed wire in 1874 led to hundreds of miles of private landholdings which enclosed vast 'cattle kingdoms' (Barclay 1980; Ingold 1980).

With the development of the new cattle ranch came the advent of the cowboy, for the boundaries had to be patrolled and some control had to be exercised over the thousands of cattle. In the beginning the man hired to watch over the cattle was considered to be part of the rancher's family group and he usually owned some cattle of his own. But, with the increased pressure on the land and the realization of the capital value of the stock, this became no longer tenable and the cowboy became a lone ranger who, on the Argentine pampas as well as on the North American plains, was often in sole charge of more than a thousand head of cattle.

The cowboy may often have been just a hired hand on horseback, and a very unromantic one at that, but as noted by Barclay (1980) the fact that he was mounted and often armed fostered his sense of equality with the rancher who employed him.

On each ranch there could be some hundreds of horses kept together in a herd, called the *remuda*, which was cared for by the young or old ranch hands. For the annual round-up of the cattle the cowboys could need between eight and fourteen horses each, so that they always had a fresh one available (Barclay 1980). The horses were highly trained to single out individual cattle and separate them from the herd, to make sharp and fast turns, and to stop and start quickly. The scarcity of trees and posts on the ranch meant that the horse could not be tied up so it was trained to stand still in one place after its rider had dismounted.

This training of the horses and the dramatic scenes which accompany the breaking-in of horses and the roping of cattle led naturally to competitions and spectacles of bravery which were the precursors of the rodeos. According to Barclay (1980) the first professional rodeo dates from 1882 when Buffalo Bill Cody put on contests in shooting, riding, and bronco-busting at the Fourth of July celebrations in Nebraska. In later years the event came to be known as *Cody's Wild West Show*.

On the South American pampas, during the late nineteenth century, the gaucho became the equivalent of the northern cowboy. Gauchos were originally of mixed race between Iberian and native American. From the eighteenth century, or earlier, they had lived as itinerant cattle herders who were expert horsemen, but as with their counterparts on the North American plains they were later forced into employment as a *vaquero* or cowboy on the vast cattle ranches. Like their forbears the gauchos often rode with a dropped noseband rather than a bit, and a toe ring for their bare feet. Their method of roping horses or cattle, as described by Azara (Hunter 1837) also differed from that in North America:

> Their mode of catching them, is, by going out in search of a wild herd, and, on arriving within cast, to throw their *bolas*, or balls, at them: these balls are three stones, about the size of a man's fist, rolled up in leather, and tied to a common centre with strong leathern cords, more than a yard in length. They take hold of one, which is rather smaller; and, after flourishing or whirling the other two several times round their head, they discharge the whole three, and entangle them in the horse's legs in such a manner, that it cannot run; affording an opportunity to throw the lasso. This lasso is a piece of cow's leather, of the thickness of a man's thumb,

very strong, with an iron bolt at the end, to make it easily cast, from twenty to thirty yards long; which they throw with admirable skill on the neck of the horse or bull they desire to catch, and detain and master him, the other extremity being affixed to the girths of their horses' saddles.

The 'rough notes of Head' which Hunter (1837) included in his translation of Azara's work on the quadrupeds of Paraguay contain a long description of the gauchos, which could rival in romanticism any account of the Hollywood cowboy:

As his constant food is beef and water, his constitution is so strong, that he is able to endure great fatigue; and the distances he will ride, and the number of hours that he will remain on horseback, would hardly be credited. The unrestrained freedom of such a life he fully appreciates; and, unacquainted with subjection of any sort, his mind is often inspired with sentiments of liberty which are as noble as they are harmless, although they of course partake of the wild habits of his life. Vain is the endeavour to explain to him the luxuries and blessings of a more civilized life; his ideas are, that the noblest effort of man is to raise himself off the ground, and ride instead of walk – that no rich garments or variety of food can atone for the want of a horse.

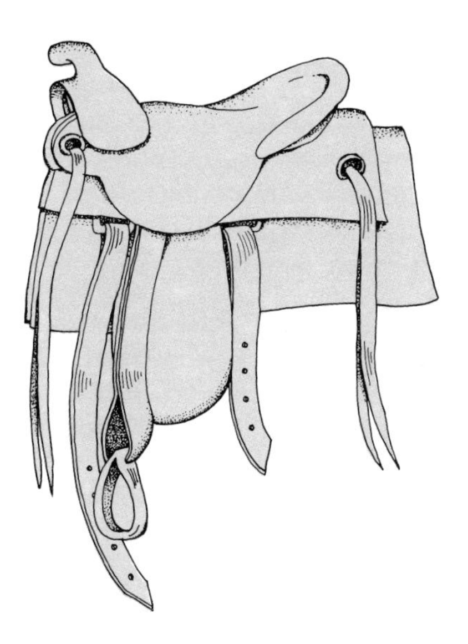

11 Equids in agriculture, transport, exploration and warfare

Animals for plough and traction AD 1100–1800

The relative importance of the horse against the ox in the agricultural systems of Europe for the seven hundred years from AD 1100 to 1800 has long been a subject of controversy. At the time of the Domesday Book (the survey of the extent, value, and ownership of the lands of England, commissioned by William the Conqueror in 1086), according to Barclay (1980) there may have been fewer than one horse for every hundred persons. Even if this figure is unrealistically low, there could only have been a rather small number of horses compared to oxen in Britain at that time. Oxen were used for agricultural work and draught, and the ordinary means of travel was by foot. The numbers of horses gradually increased until by the eighteenth century there was a patchwork of agricultural use with horses in some parts of Britain and oxen in others. However, the main purpose for which horses were kept continued to be the provision of a breeding stock for road transport and warfare.

Factors that led to the changeover from oxen to horses on the farm were the spread of the horse collar as a method of harnessing instead of the neck strap (see Chapter 5), and the improved nutrition made possible by the three-field system of crop rotation. Under the old system half the land was planted with winter grain and the other half was left fallow; the next year the planting was reversed so that the land always had a year of recovery between crops. Under the improved system, which came into widespread use in northern Europe in the Middle Ages, the arable land was divided into three parts. The first field was planted in the autumn with winter wheat or rye. The following spring the second field was planted with oats, barley or pulses, and the third field was left fallow. The next year the first field was planted with summer crops, the second field was left fallow, and the third field was sown with winter cereals. This system resulted in a much higher productivity of crops and a surplus of oats which could be fed to horses that were harnessed to the heavy plough (White 1962).

Oxen also benefited from being fed on oats but they needed only half the quantity required by horses and could thrive on much poorer pasture. On the other hand their slow pace of ploughing was a distinct disadvantage. In general the ox was preferred on lowland areas with heavy clay soils and the horse on upland areas with thin stony soils, on which plough-oxen easily slip. However, it was not only the provision of oats and the type of soil that decided whether the ox or the horse predominated. There was also the question of social traditions, which allowed a plough-ox to be slaughtered for its meat when it grew too old for work, while the eating of horse-meat, although it was occasionally practised, never became widespread. Furthermore, the farming communities did not like change, so the plough animals that were good enough for the fathers were considered good enough by the sons. Another factor was that, although landowners were anxious to see the speed of ploughing increased, farm labourers were not keen on the extra work.

In southern Europe and the Mediterranean region the donkey and mule were always favoured above the horse for agricultural work and often a donkey and an ox, or a mule and an ox were harnessed to a plough together (Fig. 11.1), despite the Biblical edict against the mixing of plough animals (see p. 94). For no very apparent reason the mule has never been favoured as a beast of burden or for ploughing in northern Europe. The donkey has been more popular but it is outside the range of its climatic tolerance and it does not flourish in the north without special attention. Even so the donkey has become adapted to the damp but relatively warm climate of Ireland, where it came into widespread use in the eighteenth century when the country was denuded of horses for the wars with France and for the army of occupation in India (Dent 1972).

'ig. 11.1 Threshing with two cows and a mule, Chios, Greece. (Photo author.)

The emergence of the heavy draught horse

Until the fifteenth century the horses of Britain were small and strong but not powerful enough to haul the increasing loads of road traffic nor to carry the great weights of armour that were demanded of them, even though, since the thirteenth century, large stallions had been imported, mostly from the Low Countries, and these had improved the stock. Trow-Smith (1957) quoted references to the import of horses to Britain from Hungary, Spain, France, Poland, Denmark, and Sweden between the fifteenth and sixteenth centuries.

Henry VIII (1491–1547), himself no light weight, was especially concerned to improve the size and strength of the horse throughout his realm. He decreed in 1535 that all substantial landowners had to keep at least two mares of 13 hands (132 cm) or over and in 1541 he prohibited the grazing of stallions of less than 15 hands (142 cm) on common land in the midland and southern counties of England (Trow-Smith 1957). These heights show how very small the ordinary horses of Britain still were in the sixteenth century.

The first horse carriages for public use were instituted in 1564 and these led to a demand for well-matched draught horses. A hundred years later the heavy horse had become established as a breed for use on the land, for haulage, and to carry the nobility in their armour (Fig. 11.2). This last was usually only on ceremonial occasions, because, with the new firearms loaded with gunpowder, heavy armour for horse and man had lost its protective value and went out of use in battle. In the seventeenth century the sturdy Medieval war-horse was replaced by lighter, more mobile cavalry horses, closer to the hunter of modern times.

At the end of the seven-year Civil War, after the Battle of Worcester in 1651, Oliver Cromwell dispensed with the last of his cavalry's heavy horses (destriers) and replaced them with as many Arab imports as he could purchase from horse buyers abroad.

Heavy draught horses need a great deal of food all year round and they could not have been bred in large numbers for farm work in Britain until the improvement of winter feed became general in the seventeenth century. The cultivation of root crops, for which 'Turnip Townshend' (1674–1738) was famed, the development of large cultivated legumes, and the improvement in grasses grown for hay all contributed to the amount of fodder available for livestock.

Working ponies and the horse trade

At the same time as the heavy horse was being developed for agricultural purposes, Scandinavia, Iceland, and the islands of Britain, where fodder was very scarce, were nurturing the small,

'ig. 11.2 War armour for a ;erman cavalry soldier and his ıorse, 1470–90. (Photo reproduced ›y permission of the Trustees, The Vallace Collection, London.)

hardy ponies that were indispensable for haulage, peat-cutting, and mining. The smallest of these, the original breed of Shetland pony was between nine and eleven hands (91.4–111.8 cm withers height) and their food was supplemented by browsing on seaweed in the winter (Trow-Smith 1959).

From Medieval times onwards there were very wide differences in the prices of horses, and as with motor cars at the present day there was a horse to suit every pocket. People became increasingly used to travelling on horseback, rather than walking, and the horse trade became a regular part of the farm economy, especially in the vicinity of cities such as London where regular horse markets were held.

Daniel Defoe wrote in 1724–6 that: 'There is also a great market, or rather fair for horses, in Smithfield every Friday in the afternoon, where very great numbers of horses, and those of the highest price, are to be sold weekly' (Rogers 1986).

Perhaps the heaviest loads that were carried on horseback or in waggons from place to place were those of the gold coinage. In Britain, paper currency did not replace gold and silver until the seventeenth century. The coins were very heavy and they had to be protected; in 1307 one thousand pounds (453.6 kg) was taken from London to Carlisle in four carts, each drawn by five horses, and there were twelve men-at-arms, sixteen archers, and a king's messenger to guard the money. Presumably each cart carried a load of around two hundred and fifty pounds (113.4 kg) of coins which seems rather low but it may of course have carried other goods as well, and the bad state of all roads made haulage by carts an inefficient form of transport.

Small amounts of goods or fresh food, such as fish, were carried by relays of pack horses, each of which could carry about two hundred pounds (90.7 kg) in weight. On the other hand two cart horses in the late fourteenth century were reputed to be able to pull a load of 2520 pounds (1143 kg), excluding the cart (Langdon 1986).

Until the coming of the railways in the 1830s transport along the coasts of Britain in ships and on canal barges was the most efficient method of moving goods around the country. This included the vast quantities of coal that were mined in the Midlands of Britain and transported south. Trevelyan (1948) claimed that twenty thousand horses were used in the transport of coal from Newcastle at the beginning of the eighteenth century; presumably this was in pulling the coal barges and in taking the loads of coal from the mines to the shipping ports.

The age of improvement

Over the course of the eighteenth century the human population of Britain probably more than doubled from around five million in 1700 to over ten million in 1800. It was a century of great change throughout Europe, of agricultural and industrial innovations and of great movements of people, none of which could have occurred without massive improvements to the roads. At the beginning of the century there were ships that could carry heavy goods to India and America, but inland the pack horse was still the usual means of transport, because the roads were so bad that wheeled traffic could make little progress. In Britain there was no highway authority and maintenance of the roads was in the hands of the local parishes

g. 11.3 A black stallion
scended from Robert Bakewell's
nproved stock of black Leicester
orses. Painted by the nineteenth
ntury artist William Shiels.

which did not care to spend money for the sake of travellers from distant parts. Private turnpike companies were therefore commissioned by Parliament to erect gates and toll bars and to exact money from users of roads in return for their upkeep (Trevelyan 1948).

The roads were slowly improved and private carriages and post-chaises came into common use. The Grand Tour of France and Italy became fashionable and everyone of note travelled abroad. The historian Edward Gibbon was told in 1758 that 40 000 English, counting masters and servants, were touring or resident on the Continent (Trevelyan 1948). However, the condition of the roads still varied greatly according to the nature of the local soils. As late as 1789 the roads of Herefordshire were impassable for wheeled carts during the winter until the end of April when the surface was levelled by means of special 'ploughs', each drawn by eight or ten horses.

Improvement of the roads had to follow from the beginnings of the industrial revolution and this encouraged the movement of large numbers of people away from the land and into the towns where there was more work. Urban life and the huge increase in the population meant that greater efforts had to be put into increasing the quantity and quality of meat and cereal foods. This was the impetus for the revolution in agriculture which paralleled the industrial revolution. *Improvement* became the great creed of the landowners and wealthy farmers, of which Robert Bakewell (1725–95) was the most notorious. He is best known for his improvements of the New Leicester sheep and the Longhorn breed of cattle but he also carried out breeding experiments with pigs and with the black horse of Leicestershire (Fig. 11.3). This breed of heavy horse was already well-defined in the 1720s at the time of Defoe's travels round Britain, when he wrote:

> The horses produced here, or rather fed here, are the largest in England, being generally the great black coach horses and dray horses, of which so great a number are continually brought up to London, that one would think so little a spot as this of Leicestershire could not be able to supply them. *Rogers 1986*

Working on this stock, Bakewell produced a horse that had a thick carcass, short straight back, and short clean legs. It was as strong as an ox, yet as active as a pony. This was the progenitor of the Midlands type of Shire horse.

Although Robert Bakewell left very few written records it is probable that he learned his methods of livestock improvement from the breeders of racehorses who had from the beginning of the century been crossing native horses with imported Barbs and Arabians (see Chapter 12).

The welfare of farm animals received little professional attention apart from the common-sense treatments of the farmer and farrier until the end of the eighteenth century. In 1791 The Veterinary College of London was founded for the study and treatment of the diseases of horses. This establishment lasted until 1872 when it was expanded to cover the treatment of all animals and was renamed The Royal Veterinary College.

Exploration

Throughout history, people have made use of animals to carry themselves and their goods around the world. The ox-waggon has been ubiquitous in the Old World; reindeer have been used in the north, camels in the deserts, the llama in South America, the dog travois in North America, but everywhere the horse and the mule have provided the most efficient means of transport and exploration.

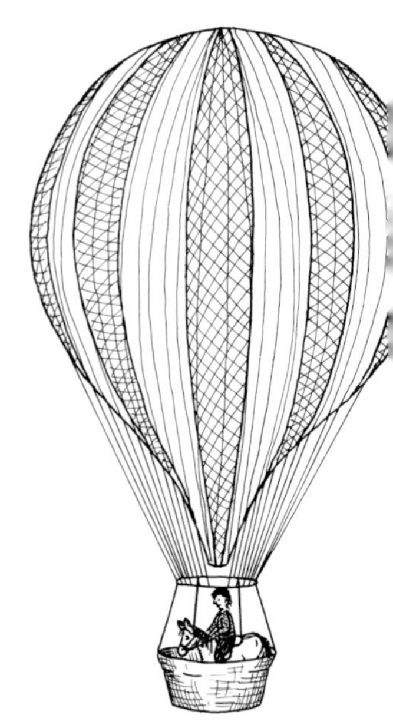

Fig. 11.4 An equestrian ascent.

The most bizarre of all equine events that could be included within the term of 'human exploration', must surely be the nineteenth century 'equestrian ascents' as described in the *Encyclopaedia Britannica* (1875, 1:192; Fig. 11.4):

> We ought also, perhaps, to notice a curious ascent made by Mr Green on July 29, 1828, from the Eagle Tavern, City road, on the back of a favourite pony. Underneath the balloon was a platform (in place of a car) containing places for the pony's feet, and some straps went loosely under his body, to prevent his lying down or moving about. Everything passed off satisfactorily, the balloon descending safely at Beckenham; the pony showed no alarm, but quietly ate some beans with which its rider supplied it in the air. Equestrian ascents have since been repeated. In 1852,

Madame Poitevin, who had made several such journeys in Paris, ascended from Cremorne Gardens, London, on horseback (as 'Europa on a bull'); but after the first journey its repetition was stopped in England by application to the police courts, as the exhibition outraged public feeling.

The most poignantly described of all explorations using horses and mules must surely be those of the British to the Antarctic at the beginning of this century (Huxley 1913). The details of Scott's last expedition and death after he reached the South Pole are very well known. The sufferings of the animals they took with them, although fully appreciated by the members of the expedition, were considered to be one of the unavoidable difficulties associated with exploration. It is reputed that the reason Scott failed was that he relied on ponies for traction and for providing meat, while the Norwegian, Amundsen succeeded because he used only dogs.

Scott and his fellow explorers left New Zealand on the ship, the *Terra Nova*, on 26 November 1910, with twenty ponies on board, together with thirty-three dogs, five tons of dog biscuits and about fifty tons of horse feed (Fig. 11.5). On Wednesday 4 January 1911, they landed and the seventeen surviving ponies were picketed on the ice flow. A stable was built for them out of the ice, and the explorers began work on unloading the goods from the ship. Not surprisingly there were many difficulties with the ponies suffering from colic, tapeworms, and many other disorders even before the southern journey began on 1 November 1911. They travelled nearly halfway

'ig. 11.5 Captain Oates with four of he ponies on the *Terra Nova* bound or the South Pole. From Huxley 1913). (Photo Mary Evans Picture library.)

to the South Pole, apparently in fairly good condition, until their fodder ran out and the last ponies were shot at Camp 31 on 9 December 1911 (Huxley 1913, 1).

In February 1912 the *Terra Nova* returned to the base at Cape Evans loaded with stores and seven mules from India, which had been ordered on the recommendation of Oates. They were used in the final expedition south which discovered the bodies of Scott, Wilson, and Bowers. Two were killed on the return journey and five survived to live in retirement at the base.

Fig. 11.6 Gran with the mule Lal Khan. From Huxley (1913). (Photo Mary Evans Picture Library.)

The mules had been very well-prepared in India for their expedition. They were equipped with canvas 'snow-goggles' and snow-shoes, and they had been exercised in 'rocking-boxes to develop the muscles especially brought into use by the motion of a ship' (Huxley 1913, 2). As would be expected, the mules were in better condition after the long sea journey, and were able to haul loads of up to 700 pounds (317.5 kg) which was far in excess of any weights hauled by the ponies in the previous year (Fig. 11.6).

British cavalry since 1800

From the beginning of the nineteenth century nearly every decade was punctuated by battles whose names became household legends, beginning with the Battle of Waterloo which was won by the British against Napoleon on 18 June 1815. In this battle it was the horses of the Household Brigade, led by Lord Uxbridge in a decisive clash against the French, that helped to win the battle. Vesey-Fitzgerald (1946) claimed that the victory was in great part due to the British horses which were heavier than those of the French, whose cavalry

Fig. 11.7 The Prince of Orange at the Battle of Waterloo, 1815. (Photo Mary Evans Picture Library.)

had been much depleted by the long period of the Napoleonic wars. As in so many battles for the millennia before Waterloo it was the side with the superior tactics of mounted shock combat that won.

The one redeeming battle for the British in the whole débâcle of the Crimean War of 1854 was the Charge of the Heavy Brigade, although it failed to receive the romantic treatment accorded to the Charge of the Light Brigade. This was one of the last battles to use horsemen in shock combat. On 24 October 1854 Lord Lucan led three hundred British horsemen into the southern valley which divided the plain at Balaclava, where they were attacked by, but managed to repulse, five thousand Russian cavalry.

In the 1870s (according to the ninth edition of the *Encyclopaedia Britannica*, II) the British cavalry consisted of thirty-one regiments which comprised the Household Brigade (the bodyguard of the Queen), seven regiments of Dragoon Guards, three of Dragoons, five of Lancers, and thirteen of Hussars.

Recruits into the Household Brigade were the tallest men (over five foot ten inches (1.78 m); they wore a helmet, cuirass, and long jack-boots, and were armed with a breech-loading carbine and long straight sword. In full dress they carried between twenty-one and twenty-two stone (133.4–140 kg). The Hussars, who were classed as light regiments, were recruited from men of five foot six inches to five foot eight inches (1.70–1.72 m). They wore blue, unlike the Dragoons who wore red, and they carried between seventeen and a half to eighteen and a half stone (111–117 kg). The Lancers also

Fig. 11.8 R.S.P.C.A. Animal War Memorial, Kilburn Park, London. (Photo The Natural History Museum, London.)

wore blue, with the Lancer's square-topped cap, and they carried a nine foot (2.74 m) bamboo lance, sword, and pistol.

A regiment of cavalry, ready for war as recently as a hundred and twenty years ago, consisted of twenty-seven officers, six hundred and seven men, with five hundred and fifty-nine horses. This included eight trumpeters, twenty farriers, shoeing smiths, and saddlers. They took with them in the field of battle two forge-waggons, one small-arm ammunition cart, and seven baggage and store waggons.

The cavalry horses were mostly obtained from dealers at the headquarters of the regiments. They were usually bought at four years old for a maximum government price of £40; the average height of the horses was fifteen and a half hands (157 cm). The veterinary surgeons attached to the regiments wore the uniform of the corps, with a cocked hat and a red plume.

During the first three years of World War I the numbers of horses and mules in the British army was increased from about 23 000 to more than a million (Vesey-Fitzgerald 1946), but this was the last time that animals played a major role in battle. The mule was replaced by the army lorry and the horse was replaced by mechanization which has progressed from the tank in 1917 to the nuclear war-heads of today. Perhaps the only benefit to come from the modern machinery of war is the reprieve it has brought to the millions of animals, sacrificed throughout history, in battles of ever-increasing ferocity which have been fought over the invasion and defence of territory (Fig. 11.8).

12 A history of horse-racing

The Homeric chariot race

Approach the goal very closely, drive your chariot and horses near; but bend a little towards the left side of the horses from your well-joined chariot seat; and cheering on the right hand horse, whip him and give him the rein with your hands. Let your left hand horse move close to the goal, so that the nave of the well-made wheel may seem to touch the goal post, but avoid touching the stone, lest you wound your horses and break your chariot in pieces, which would be a joy to others and a disgrace to yourself.

Iliad 23, 331–360*

Thus Nestor advised his son, Antilochus, on how to drive round the goal post in the earliest known description of a chariot race. It is recounted in the Iliad, a Homeric poem about the Trojan War that took place at the end of the thirteenth century BC. It is generally believed that the Iliad was composed during the second half of the eighth century BC, that is about 500 years after the fall of Troy.

The chariot race, as described by Homer, was a central event in the funeral games for Patroklos (the close friend of Achilleus) who was killed during the Trojan War. His body was recovered by Achilleus and burned in the centre of a pyre, a hundred foot wide, together with the sacrifice of four of his own horses and two of his nine dogs, as well as twelve young Trojan men, and a large number of cattle and sheep. After the cremation the white ashy bones of Patroklos were extracted from the pyre and placed in a gold vessel, covered with a double layer of animal fat and interred in an earth mound.

Achilleus then set out the five splendid prizes for the winners of the chariot race. The first prize was a faultless woman and a handled tripod, holding twenty-two measures. The second prize was an unbroken six-year-old mare, pregnant with a mule foal. The third prize was a shining metal tripod, holding four measures. The fourth was two talents of gold and the fifth was a double vessel untouched by the fire.

The race was run across the open plain, outside the city of Troy, to a goal and back again to where Achilleus would judge the winners. The goal was a standing wooden post of oak or larch (perhaps a monument to some man long since dead) with two white stones on either side of it.

This and the following passages are aken from the translations of uckley (1853) and Hammond 1987).

The contestors were:

Eumelos, son of Admetos in Thessaly, and owner of the fleetest horses in the Greek army. He drove a pair of mares.

Diomedes, son of Tydeus and one of the bravest of the Greeks in the Trojan War. He drove a pair of stallions captured from Tros, son of Erichthonios, king of Troy.

Menelaos, son of Atreus, who drove the mare, Aithe, yoked to his own stallion, Podargos.

Antilochus, son of Nestor, who drove Pelian-born stallions that were not very fast. Therefore Nestor exhorted Antilochus to use all his cunning to drive his chariot more skilfully than the others.

Meriones, a charioteer of Idomeneus, King of Crete, who was to watch the race from a high point outside the circus. He was the least skilfull driver and his stallions were the slowest.

When all the charioteers were ready they cast lots into a helmet which was shaken by Achilleus. Antilochus' lot was drawn first, followed by those of Eumelos, Menelaos, Meriones, and Diomedes. They lined up in that order, then all raised their whips at the same moment and struck their horses with the reins:

> They rapidly flew over the plain, far away from the ships, swiftly, and beneath their breasts the excited dust stood up, raised like a cloud or a whirlwind; whilst their manes were tossed about by the breath of the wind. Sometimes, indeed, the chariots approached the fruitful earth, and at others bounded aloft; but the drivers stood erect in the chariots, and the heart of each of them, eager for victory, palpitated.

It was not, however, a straight gallop; Eumelos fell out of his chariot, the yoke was broken, the pole dashed to the ground, and his mares ran out at either side. He, himself, was thrown from the driving platform and was lacerated around his arms, mouth and nose, and his forehead was bruised near his eyebrows. Meanwhile the watching Cretan, Idomeneus, began to shout and give a running commentary of the race which incensed the other spectators, so that Aias told him to stop prating, and Achilleus had to intervene and put a stop to their angry discourse.

The first to complete the course to the goal and back again was Diomedes who lept from his chariot in the centre of the circus. He received the first prize of the woman and the tripod and gave them to Sthenelos, king of Mycenae, while he unyoked his horses which were streaming with sweat. Antilochus drew up next, followed by Menelaos and then by Meriones. Eumelos was last and he came in dragging his chariot and driving his mares ahead of him.

Then came an argument about how the next prizes should be allotted. Achilleus felt sorry for Eumelos and wished to give him the second prize, and Menelaos accused Antilochus of cutting in ahead of him and preventing him from winning. Antilochus apologized and was forgiven, and Eumelos received in consolation, instead of the mare (which was the second prize), a valuable bronze and tin corslet from Achilleus' house.

Although this event is supposed to have taken place five hundred years before the time of Homer, the account is so detailed and so vivid that it must surely either have been passed down unchanged through the centuries as oral poetry, or be the description of a race that Homer, himself, had watched. Besides being a seemingly accurate account of how chariots were driven, details are given of the horses which are of considerable interest. First, it was not only stallions which were yoked to the chariots but mares were also used and one pair even consisted of a stallion and a mare. It is no wonder that there were only five contesters in the race, with each deserving a prize, when it is considered what skill would be required to drive a wooden chariot at full gallop across a sandy plain, yoked by a neck collar and pole to a stallion and a mare.

Second, there is the description of the manes of some horses, which are said, 'to hang to the ground'. This may reflect the fact that in the wild horse and probably also in early domestic horses the mane was normally short and upstanding. A hanging mane is a feature of domestication which, although seen in depictions of horses from as early as the second millennium BC, may still have been rather uncommon in Homer's time. Chariot horses were usually shown with cropped manes, presumably to prevent the hair becoming entangled with the reins. Third, Homer describes the colour of one horse as being *phoinix* which is usually translated as chestnut, all over except for a round white spot, like the moon, on its forehead. The horses of Meriones had beautiful fair manes which may have meant that they were hanging and light-coloured.

'ig. 12.1 Four-horse chariot race on the shoulder panel of a sixth century BC Attic black figure vase. Photo reproduced by courtesy of the Trustees of the British Museum.)

The Olympic Games

Legend holds that the games held at Olympia in ancient Greece were instituted by Zeus after his victory over the Titans in 1453 BC or by Hercules in honour of Olympian Zeus in 1222 BC. The games did not,

however, achieve outside fame until Coroebus obtained the prize in 776 BC. This was counted as the first Olympiad and from then on the space of time between each occurrence of the games was called an Olympiad. The games were held once every four years at the time of the first full moon after the summer solstice. Dating by means of the Olympiads was popular in literary works until the end of the fourth century AD when it was forbidden under the rule of the Roman emperor Theodosius the Great, who died in AD 395. Four-horse chariot races were introduced to the games at the 23rd Olympiad (684 BC) and mounted horsemen competed in races at the 33rd (644 BC) (Figs. 12.1–12.3).

Fig. 12.2 Harnessing a four-horse racing chariot. On the same Attic vase as Figure 12.1.

Fig. 12.3 Two-horse chariot race on an Attic red figure vase. Mid-fifth century BC. (Photo reproduced by courtesy of the Trustees of the British Museum.)

The Roman chariot race

The four-horse chariot or *quadriga* played an essential role in the Roman circuses where all races were run, while other athletics, animal spectacles, baiting, and competitive sports took place in the amphitheatres and stadia. The circus was an oblong or oval enclosure with rising tiers of seats around it (Fig. 12.4). Every major city had at least one and Rome had five, with the huge Circus Maximus being the largest. The arena of this circus, which according to Pliny could hold 260 000 spectators, was 2150 feet (650 m) long by 725 feet (220 m) wide and it had a central spine, around which the horses had to turn, of 770 feet (233 m) length.

Each chariot race was normally seven laps which meant, according to Hyland (1990), that the horses galloped a distance of about three miles (4.5 km). The surface of the arena was made of compacted sand. As with the Greek chariots, the two central horses were harnessed to the pole and the two outer horses were attached by traces. The chariot driver stood on a simple box between the two wheels and must have needed very great skill to manage the horses and to turn them sharply round the spine in the centre of the arena. The starting line was painted white and there was often a system of starting stalls or *carceres* which were operated with a system of levers so that they all opened at once (Hyland 1990).

The rituals of the chariot race were just as elaborate as those of the modern horse race. The charioteers of Rome were divided into four factions (companies), whose drivers competed against each other, as mentioned by Pliny in his *Natural History* (Book 7), written during the first century AD. Each faction was distinguished by a different colour and dedicated to a season of the year. There was green for spring, red for summer, blue for autumn, and white for winter.

There are remarkably detailed records of the chariot races and their drivers. Hyland (1990) quotes the Latin inscription about the charioteer Diocles in the mid-second century AD. From the age of eighteen in AD 122, for a period of twenty-four years, Diocles drove 4257 races with 1462 wins. His best races were for the red faction, but like all charioteers, most of whom were slaves, Diocles passed from faction to faction. Ownership of the horses was under the control of the factions and was established by branding.

ig. 12.4 The Lyon Circus mosaic howing four-horse chariots, the tarting stalls, and the white lines for tarting and finishing the race. Photo Musée de la Civilisation Gallo-Romaine.)

The great popularity of the chariot races throughout the Roman Empire meant that there was a constant demand for horses which caused a depletion in the supply available for the cavalry. This, together with the sometimes extremely unruly behaviour of the spectators, led to there being much pressure on the state to control the sport.

The chariot horses were nearly always stallions whose training began when they were three years old, but they were not raced in the circus until they were five years old (Hyland 1990). Although the four-horse chariot seems to have been the most popular for racing, there are also records of two-horse chariots and of races with mounted jockeys (Fig. 12.5).

Fig. 12.5 Four named racing stallions with their jockeys. Mosaic in the Musée de Sousse. From Hyland (1990).

As in the modern era of horse-racing the fastest horses seem to have been those, presumably of Arabian or Barb conformation, that were imported to Rome by ship from North Africa.

Horse-racing in Britain from AD 1100 to Eclipse

The first lengthy account of horse-racing in British history is that of William Fitzstephen in his *Description of the City of London*, written in Latin *c.* 1174. The following translation of his account of racing near Smithfield Market is given in Youatt (1846):

> Three jockeys, or sometimes only two, as the match is made, prepare themselves for the contest. The horses on their part are not without emulation: they tremble and are impatient, and are continually in motion. At last, the signal once given, they start, devour the course, and hurry along with unremitting swiftness. The jockeys, inspired with the thought of applause and the hope of victory, clap spurs to their willing horses, brandish their whips, and cheer them with their cries.

During the time of the Crusades, tournaments and jousting on war horses was more popular than the straight race. However, in the reign of Richard I, knights rode at Whitsuntide over a three-mile course for forty pounds of gold. In 1512 public races were established at Chester and in 1540 a silver bell, valued at three shillings and sixpence or more was given, 'to him who shall run best and furthest on horseback before them [the Company of Saddlers and the Mayor of Chester] on Shrove Tuesday; these bells were denominated St George's bells'.

During the reign of James VI (1566–1625) of Scotland and I of England race meetings were established near Richmond in Yorkshire, and at Croydon and Enfield Chase. These were attended by the King who also went to races at Epsom and he built a house at Newmarket, where there is a record of horse-racing for 1605. Charles I (reigned 1625–49) continued the patronage of racing, especially at Newmarket where annual races have been held since 1667.

In 1654 and 1655 Cromwell forbade horse-racing for some months (presumably because of its immoral effects on the populace), but after the restoration in 1660 Charles II provided a new impetus for the sport and rebuilt the house of James I at Newmarket which had fallen into decay. Racing came under new regulations of weights and distances and Charles II was the first monarch to run horses under his own name.

It is to Queen Anne (who reigned from 1702–14) and her consort George, Prince of Denmark, that the origins of modern racing and the Thoroughbred racehorse can be attributed. The Queen and Prince George were zealous patrons of the turf and actively pursued a policy of improvement by importing Arab stock, including the Darley Arabian, one of the most influential of all Arabians (Youatt 1846). In 1703 races were established at Doncaster and in 1709 at York where Queen Anne's gelding Pepper ran for the Royal Cup in 1712.

In 1739, during the reign of George II, an act was passed to prevent racing by ponies and weak horses. The Jockey Club was founded in 1750 and purchased the racing ground at Newmarket in 1753. In 1776 Richard Tattersall established Tattersalls, the famous centre for betting and for the sale of horses at Hyde Park Corner in London, where it remained until 1865.

From this time, flat-racing became a national sport; the St Leger was established in 1776, the Oaks in 1779 and the Derby in 1780. Steeplechasing originated in 1793 as a cross-country race in which a church steeple was the goal, and all intervening obstacles had to be jumped in order to reach it. The Liverpool Grand National, run each year at Aintree, dates from 1839 when it was won by a horse called Lottery.

Influence of the Arab and the early Thoroughbreds

It seems that, in Britain, there was always a tradition that horses from the East were superior to those of native origin. Indeed it is to be expected that the long-legged more gracile horses from the hot deserts of Arabia and North Africa would gallop faster than the stocky ponies of the north.

The first written evidence for the import of an Arabian horse into Britain was in AD 1121, when Alexander I King of Scotland presented to the church of St Andrew's, 'an Arabian horse with costly furniture, Turkish armour, many valuable trinkets, and a considerable estate' (Youatt 1846). From that time onwards, Turkish and Barbary horses were regularly imported in efforts to improve the speed of native horses. James I paid five hundred pounds to a Mr Markham for a celebrated Arabian horse, but it met with disfavour from the Duke of Newcastle who wrote a book on horsemanship and described this horse as a little bony pony of ordinary shape. After this the Arab went out of favour with English breeders until Charles II sent his master of the horse to the Levant to purchase brood mares and stallions. Not all these horses came from Arabia, however, for one at least of the royal mares was a Barb, the dam of the famous horse Dodsworth.

It should be noted here that, while the Arabian horse originates from western Asia, the Barbary or Barb comes from the coastal belt of north-west Africa in the countries of Morocco, Algeria, and Tunisia. Physical characteristics that distinguish the pure Barb from the Arab is its slight Roman or 'ram-like' profile to the facial region, a sloping croup (rump), and a low-set tail. A Thoroughbred is any horse whose pedigree is recorded in the *Stud Book*. The first attempt to record pedigrees of horses in England was in 1791 when an *Introduction to a General Stud Book* was published by Weatherby's, the official agents of the Jockey Club. The first volume of the *Stud Book*, also published by Weatherby's, appeared in 1808 and has continued to be produced periodically since then.

As is well known, all modern racehorses and Thoroughbreds, cited in the *Stud Book*, are descended from three Arabian stallions, the Byerley Turk, the Darley Arabian, and the Godolphin Arabian. Little is known of the Byerley Turk (Fig. 12.6 see p. 125), the first of these three, except that, in the words of the *Stud Book*, 'he was Captain Byerley's charger in Ireland in King William's wars', which would have been in about 1689. Byerley Turk did not sire many well-known horses except Jig in about 1705, who was the sire of Partner (Haralambos 1990).

The import of the Darley Arabian from Aleppo at the beginning of the reign of Queen Anne was the start of a new era in horse-breeding

(Fig. 12.7). According to the *Stud Book*, 'Darley's Arabian was brought over by a brother of Mr Darley of Yorkshire, who, being an agent in merchandise abroad, became member of a hunting club, by which means he acquired interest to procure this horse'. The Darley Arabian was the most influential of the founders and his direct descendants were the Devonshire Flying Childers (so-called because he was bred by a Mr Childers and sold to the Duke of Devonshire) and the Bleeding or Bartlett's Childers, the great grandsire of Eclipse (Youatt 1846).

The story of the Godolphin Arabian, who was more probably a Barb, is intriguing. He was a brown bay, about 15 hands (152.4 cm)

g. 12.7 The Darley Arabian. After painting by John Wootton (Photo res Ltd., E T Archive.)

Fig. 12.8 The Godolphin Arabian. Engraving by George Stubbs. (© B Museum, Photo E T Archive.)

in stature, with a beautiful head, but an unnaturally high crest (Fig. 12.8). The horse was imported by Mr Coke from Paris where he had been found drawing a cart. Mr Coke gave him to a Mr Williams who, in turn, presented the horse to the Earl of Godolphin. His value was unknown until 1731 when he was being used as a teaser on the mare Roxana who was to be mated with Hobgoblin. However, because this stallion refused to have anything to do with Roxana the Godolphin Arabian was allowed to mate with her. The resulting foal was Lath who became the most celebrated racehorse of his day after Flying Childers, and the beginning of a dynasty which Youatt (1846) saw as even more successful than that of the Darley Arabian.

Eclipse (Fig. 12.9), the great-grandson of the Darley Arabian, is probably the best known racehorse of all time. His sire was Marsk, his dam Spiletta, and he was born on 1 April 1764 during an eclipse of the sun. He was a chestnut with a white blaze down his face; his lower right hind leg was white and he had black spots on his rump, which were said to be inherited by generations of his male offspring. According to Youatt (1846) Eclipse was a 'thick-winded horse, and puffed and roared so as to be heard at a considerable distance'. He was bred by the Duke of Cumberland and sold, on the Duke's death, to Mr Wildman, a sheep salesman, for seventy-five guineas. Colonel O'Kelly at first bought a half-share in the horse from Wildman for 650 guineas, but later bought him out for a further 1100 guineas. Eclipse only ran for two seasons, beginning in 1769 when he was five years old (Ulbrich 1986).

g. 12.9 Eclipse painted by George ubbs in 1770. (Photo Jockey Club, ewmarket, England.)

It is recorded by Youatt that some persons who came late to watch Eclipse in a trial asked an old woman if there had been a race. She replied that she could not tell whether it was a race or not but that she had just seen a horse with a white leg running away at a monstrous rate, and another horse a great way behind, trying to run after him; but she was sure he never would catch the white-legged horse if he ran to the world's end. From then on it was 'Eclipse first, and the rest nowhere!'. In the two seasons he ran 21 races and was never beaten. His last race was for the King's Plate at Newmarket on 18 October 1770, after which he was used as a stallion. He produced 334 offspring that were winners, and made £25 000 for his owner in service fees. Eclipse died on 25 February 1789 at the age of twenty-five years, and from then on he became a legend.

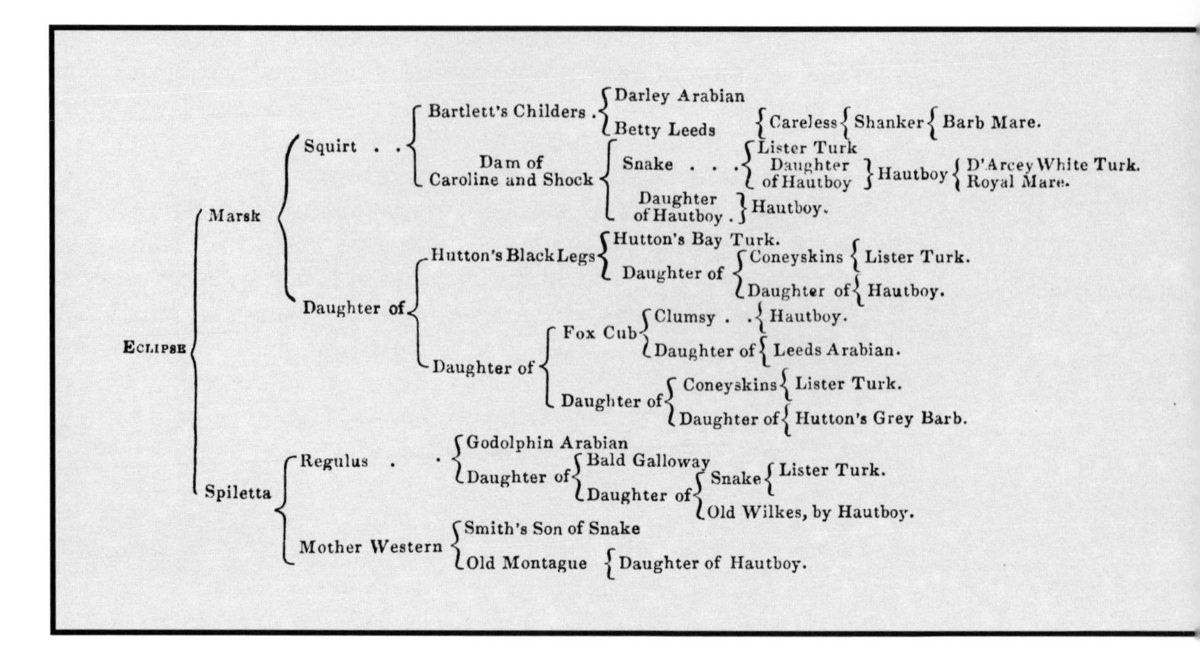

Fig. 12.10 The pedigree of Eclipse From Youatt (1846).

The pedigree of Eclipse (Fig. 12.10) is of great interest because it shows how widely the imported Arabians and Barbs were being used in attempts to improve the native stock of racehorses. The lifetime of Eclipse spanned the early part of the agricultural revolution which paralleled the industrial revolution of the eighteenth century. The rapid increase in the human population and the development of cities at this time necessitated the great boom in livestock improvement which is epitomized by the 'in-and-in' breeding experiments of Robert Bakewell who lived from 1725–95 (see Chapter 11).

It must be remembered that the livestock improvers were working in a period that was nearly a hundred years before Darwin put forward his theory on the origin of species by natural selection (1858). Breeds of dogs, for example, must have been produced by selective breeding for at least two thousand years but, although some farmers had practical experience of how to achieve inheritance of favoured characteristics in their stock, they had had no knowledge of its mechanism. Before the time of Bakewell and his followers it was believed that the way to obtain higher productivity from the animals was to feed them better, and it was a temptation to send the best animals to market and retain the poorest for breeding. The recording of the pedigrees of racehorses and the selection of sires of outstanding performance provided the paradigm for the general realization that quality in the parents could be inherited by the offspring, and this changed the whole basis of livestock breeding in the eighteenth century (Russell 1986).

It was his observations on the breeding of domestic animals and the writings of Garrard, Marshall, Young, Low, and Youatt, amongst others, that first impressed Charles Darwin and provided the impetus for his investigations into the process of evolution. Darwin was intensely interested in all forms of variation and improvement within the plant and animal kingdoms and, without any knowledge of genetics, he realized that there must be a point beyond which improvement in one characteristic, such as speed, could not be extended. He declared that:

> He would be a bold man who would assert that perfection has been reached. Eclipse perhaps may never be beaten until all race-horses have been rendered swifter, through the selection of the best horses during many generations; and then the old Eclipse may possibly be eclipsed; but as Mr Wallace has remarked, there must be an ultimate limit to the fleetness of every animal, whether under nature or domestication; and with the horse this limit has perhaps been reached. *Darwin 1868, 2*

Today, the owners of racehorses are concerned at the lack of increase in speed over the past fifty years, and Darwin's question whether the ultimate improvement has been reached is being asked, not only by racing enthusiasts, but also by geneticists. Today's Thoroughbreds are all descended from the thirty-one horses that were registered in the *Stud Book* in 1791, and one theory is that racehorses are becoming too inbred (Hill 1988). According to Gaffney & Cunningham (1988), however, the modern populations of Thoroughbreds still retain untapped genetic potential to run faster. Whichever of these theories is correct, as Dunbar (1985) wrote: 'Breeding racehorses is an expensive business that is unusually subject to luck, and fortunes will continue to be won and lost on the whinny of a highly strung stallion'.

Conclusions

The family Equidae is unusual amongst mammals in that all the species of horse, ass, onager, and zebra can be interbred and will produce hybrid offspring that are physically normal, although almost always infertile. Most taxonomists now include all the equids in one genus, *Equus* (see Appendix), and cytogenetic studies indicate that, although there are wide variations in the chromosome numbers (see p. 19), the genetic distances between the different species are low (Power 1990). Despite this taxonomic congruity, and behavioural similarities between the species, only two wild equids have been domesticated. These are the horse (*Equus ferus*) from the grasslands of northern Eurasia, and the ass (*Equus africanus*) which evolved in the hot deserts of North Africa and Arabia.

Domestication of these two equids has led to their diffusion thousands of miles from their natural species boundaries and into environments which would appear to be totally unsuitable for their natural adaptations. The transformation of the wild ass, a true desert animal, into the short-legged, long-coated donkey living in the damp fields of Ireland, more than a thousand miles north of its natural range, must be one of the most remarkable examples of evolution within a domestic species.

It is a notable fact that despite thousands of years of environmental interference and interbreeding with domestic stock neither Przewalski's horse, which is closely related to the ancestor of the domestic horse, nor the wild African ass, progenitor of the donkey, has yet become extinct. This is especially remarkable in the light of the extermination of the wild progenitor of domestic cattle, the aurochs, *Bos primigenius*, which appears to have been more widespread and more adaptable than the horse in the early Holocene of the northern hemisphere.

The Przewalski horse (*Equus ferus przewalskii*) has been extinct in the wild since the 1960s but has bred well in captivity. In 1991 an area of land in the Gobi Altai was designated as a world heritage site where captive-bred horses will be set free in their country of origin. For the first ten years of this project, organized by the Przewalski Horse Global Management Plan Working Group, ten captive-bred stallions and eight mares will be kept in a fenced area of 10 000 hectares (Sattaur 1991). When the population has grown to a viable number the horses will be allowed to roam freely over the Mongolian steppes as, until the middle of this century, they had done for a million years.

There may be a few wild asses (*Equus africanus somaliensis*) surviving in the eastern Sahara and there is a small breeding group in a wildlife reserve in Israel, as well as a few in zoos around the world.

That the wild horse and wild ass have survived until the present is probably due to the ability of these equids to retreat into hardship zones where the human population is in very low numbers. However, other species of equid have been exterminated by merciless hunting, for example the now extinct quagga (*Equus quagga*) which Harris described in 1839 as: 'Still found within the Cape colony. Inhabits the open plains south of the Vaal River in immense herds'. In Asia there was a tradition of running down wild onagers with relays of horses which lasted from the time of Xenophon to that of Tegetmeier & Sutherland (1895). The Mesopotamian onager (*Equus hemionus hemippus*) was hunted throughout history but, like the quagga, it only became extinct at the end of the last century.

The wild horse and ass should have a special value not only for conservationists of wildlife but also for the breeders of livestock, for these equids provide a store of ancestral genetic material, from which all the domestic breeds have been developed by natural and artificial selection, and which once lost, can never be recovered.

Since the beginnings of their domestication the horse and the donkey have, in one important respect, nearly always been perceived as different from other livestock: they are treated as individual animals rather than as part of a herd or a flock. Like a dog, a horse or a donkey is a partner and therefore not to be killed merely for food. It is only when the animal loses this individuality that it can be regarded as a supplier of meat, as dogs are in China or horses in France.

The special place held by the horse in human societies for at least the past four thousand years is reflected in the numerous sacrificial burials of horses that are found on archaeological sites throughout the ancient world. These sacrifices reached a peak of elaborate ritual in the tombs of the Scythian nomads whose culture was centred on the horse during the first millennium BC. Life in the next world could not be contemplated without material possessions, and these included large numbers of horses which were sacrificed and buried with their splendid trappings. The worlds of the gods differed little from the living world, as vividly described by Homer some hundreds of years before the period of the elaborate Scythian tombs:

> There is a wide cave, deep down in the water, half-way between Tenedos and rocky Imbros. There Poseidon the earthshaker reined in his horses,

unyoked them from the chariot, and tossed immortal fodder down for them to eat. And he put golden hobbles round their feet, which could not be broken or slipped, so they should wait there unmoving for the return of their master. *Iliad 13, 35, Hammond 1987*

It was not only the Scythians who sacrificed horses on a grand scale, for all over the Celtic world the horse appears to have been treated as worthy of divine attention. In Germany, eastern Gaul and Britain representations of the goddess Epona have been found in which she is shown either riding a horse or with horses on either side of her (Davidson 1988). In Norway, sacred horses were kept in the sanctuary of the goddess Freyr at Trondheim and in the sagas there are descriptions of horses kept near Freyr's temples in Iceland. Belief was widespread that sacred horses could understand the will of the gods, and also that a horse could carry a dead hero into the next world (Davidson 1964).

Throughout history, in the world of the living, the horse has always been a powerful symbol both for fertility and for warfare. Since ancient Egyptian times paintings and sculptures have been made of individual horses, while in literature the horse has been used to boost the image of the warrior and hero, and never more eloquently than by Shakespeare:

I saw young Harry, with his beaver on,
His cushes on his thighs, gallantly arm'd,
Rise from the ground like feather'd Mercury,
And vaulted with such ease into his seat
As if an angel dropp'd down from the clouds
To turn and wind a fiery Pegasus
And witch the world with noble horsemanship.
Henry IV Pt. 1, Act IV, Sc. 1

Or the horse itself may be extolled as an ethereal being that will protect the rider, as in the Dauphin's speech before the Battle of Agincourt in Henry V (Act III, Sc. 7):

le cheval volant, the Pegasus, *qui a les narines de feu*! When I bestride him, I soar, I am a hawk: he trots the air; the earth sings when he touches it; the basest horn of his hoof is more musical than the pipe of Hermes . . .

[Interjection by Orleans] He's of the colour of the nutmeg.

[Danph.] And of the heat of ginger. It is a beast for Perseus: he is pure air and fire; and the dull elements of earth and water never appear in him but only in patient stillness while his rider mounts him: he is indeed a horse . . .

In their attitudes to their animals humans have always been as confused as they are in any other relationship, mixing compassion with cruelty and altruism with commercial greed. Great numbers of horses, reared either in preparation for their sacred journey into the next world, or for warfare, were presumably always well fed and cared for. Even greater numbers of ordinary horses, donkeys, and mules that were used for draught and travel in the ancient world, as in more recent times, had short lives, during which they were subjected to starvation and unbearable hardship.

For good and bad, the horse, donkey, and mule have been companions to humans in every walk of life. Until the machine age, their services were required for almost every endeavour, and in many of these, the horse has been an unwitting accomplice in the destruction of the natural world and the needless slaughter of untold numbers of wild animals. Not least, in this destruction, must be counted the depopulation, in Australia and the Americas, of the aboriginal inhabitants who had lived in balance with their environments for many thousands of years before the European invasions.

The devastating effects of 'the discovery of the New World' were, in great part, brought about by the ability of the invaders, in the beginning, to travel fast on horseback and, later, to transport heavy loads by mule train. As reviewed by Zubrow (1990), the rapid spread of European diseases resulted in the decimation of the native Americans, probably as much from collapse of the sociopolitical systems of tribal groups as much as from the actual epidemics.

It is time for the realization that there was no discovery of a New World, only the invasion of two continents that had their own, long-established, indigenous human cultures. Without the horse and without pack animals this invasion could not have succeeded. Today, in this decade of the five hundredth anniversary of the landing of Christopher Columbus in 1492, there is a great need for a wider understanding of the richness of past cultures. An even greater need is for a new world strategy for the conservation of all life on Earth and the suppression of cruelty to both humans and animals.

Appendix
Nomenclature of the family Equidae as used in this book

The international rules for giving Latin, or binomial, names to animals and plants follow the conventions used by Linnaeus in his *Systema naturae*, published in 1758. In this work every species was given a genus name, for example *Equus*, and a species name, for example *ferus*. Later, naturalists began to use the trinomial which may be added to denote a subspecies, for example *Equus ferus przewalskii*.

The classic definition of a species is that it is a group of actually or potentially interbreeding natural populations, which is reproductively isolated from other such groups. A subspecies is a geographic population which differs in appearance (phenotypically) from other geographic populations (or races) within the range of the species.

Linnaeus classified the genus *Equus* into three species: *Equus caballus*, the European horse; *Equus asinus*, the ass (donkey), onager, mule, and hinny; and *Equus zebra*, the zebra. Both the wild horse and the domestic horse would be placed within his *Equus caballus*, and both the wild and the domestic ass within his *Equus asinus*. Linnaeus did not realize that the African and Asiatic asses are separate species that when mated will produce offspring which, like the mule, are infertile.

A.1

There are problems in having the same name for both the wild and the domestic form of a species (Clutton-Brock 1987). One difficulty is that more than one wild subspecies or even species can contribute to the ancestry of the domestic form. Both the horse and the donkey could be descended from more than one subspecies of *Equus ferus* and *Equus africanus*.

Domesticated animals are normally bred in reproductive isolation from their wild progenitors and, as a result of artificial selection, look very different from them, although they remain fully fertile when mated with the wild species. Domestic animals are not therefore genetically separate from the wild species, but they do require different names which are consistent and which by tradition and common usage are known to everyone. (It would be confusing, for example, if the dog, *Canis familiaris*, had the same name as its progenitor the wolf, *Canis lupus*.) The most practical solution is to call the domestic form by the oldest name, which is usually that given to it by Linnaeus, and to use the next oldest Latin name for the wild species. Following this system for the classification of the Equidae (or horse family), the wild species that have inhabited Europe, Asia, and Africa during the Holocene (the past 10 000 years) and their domestic derivatives are listed as follows*:

The species of equid that became xtinct in the early Holocene (10 000 ears ago) of North and South merica are not included because heir taxonomy is little understood nd, as fossils, they lie outside the cope of this book.

DOMESTIC FORM	WILD SPECIES
Equus caballus L. Domestic horse	*Equus ferus ferus* Boddaert, 1785 Tarpan or Eurasian wild horse (extinct) *Equus ferus przewalskii* Poliakof, 1881 Przewalski's (Mongolian) wild horse (extinct in the wild)
Equus asinus L. Domestic ass/donkey	*Equus africanus africanus* Fitzinger, 1857 Nubian wild ass (extinct) *Equus africanus somaliensis* Noack, 1884 (This subspecies name has priority over *Asinus somalicus* Sclater, 1884) Somali wild ass (endangered**) *Equus africanus 'atlanticus'* P. Thomas, 1884 Algerian wild ass (extinct) *Equus africanus taeniopus* Heuglin, 1861 Heuglin's wild ass (extinct, unless it was a population of feral asses) *Equus africanus dianae* Dollman, 1935 (probably a population of feral asses)

Equus hemionus hemionus Pallas, 1775
Chigetai, Kulan, or Mongolian wild ass (vulnerable**)
Equus hemionus luteus Matschie, 1911 (see Groves 1974)
Gobi dziggetai (not in the IUCN *Red Data Book*)
Equus hemionus onager Boddaert, 1785
Persian onager or Ghor-khar (vulnerable**)
Equus hemionus khur Lesson, 1827
Indian wild ass or khur (endangered**)
Equus hemionus hemippus I. Geoffroy, 1855
Syrian wild ass or Achdari (extinct)
Equus kiang Moorcroft, 1841
Kiang of Tibet (vulnerable**)
Equus hydruntinus Regalia, 1904
Wild 'ass' of southern Europe and western Asia (extinct in the early Holocene)
Equus grevyi Oustalet, 1882. (Fig. A.1).
Grevy's zebra (vulnerable**)
Equus burchelli Gray, 1824. (Fig. A.2).
Burchell's or common zebra
Equus zebra L.
Mountain zebra (endangered**)
Equus quagga Gmelin, 1788. (Fig. A.3).
Quagga (extinct)

* Status as defined in the *1988 IUCN d list of threatened animals.* ambridge: IUCN Conservation Ionitoring Centre.

hese illustrations are from the riginal descriptions of A.1 Grevy's nd A.2 Burchell's zebras, and A.3 ne Quagga.

.2

A.3

References and publications for further reading

ANTHONY, D.W. 1986. The 'Kurgan Culture,' Indo-European origins, and the domestication of the horse: a reconsideration. *Current Anthropology* 27, 4, 291–313.

ANTHONY, D.W. 1991. The domestication of the horse. In *Equids in the ancient world* II. R. H. Meadow & H.-P. Uerpmann (Eds). Wiesbaden: Dr Ludwig Reichert.

ANTHONY, D.W. & D.R. BROWN 1991. The origins of horseback riding. *Antiquity* **65**, 22–38.

ANTONIUS, O. 1937. On the geographical distribution in former times and today of the Recent Equidae. *Proceedings Zoological Society of London* **107**, Ser. B, 557–564.

ANTONIUS, O. 1944. Beobachtungen an Einhufern in Schönbrunn XVII. Halbeselbastarde. *Der Zoologische Garten* (N.F,) **16**, 1/2, 1–14.

AZZAROLI, A. 1985. *An early history of horsemanship.* Leiden: E.J.Brill.

BAHN, P.G. 1978. The 'unacceptable face' of the West European Upper Palaeolithic. *Antiquity* **52**, 183–192.

BAILLIE-GROHMAN, W.A. & F. (Eds.) 1904. *The master of game by Edward, second Duke of York.* London: Ballantyne, Hanson & Co.

BARCLAY, H.B. 1980. *The role of the horse in man's culture.* London: J.A. Allen.

BERGER, J. 1986. *Wild horses of the Great Basin social competition and population size.* Chicago & London: The University of Chicago Press.

BODDAERT, P. 1785. *Elenchus animalium* 1, *160.*

BOESSNECK, J. 1970. Ein altagyptisches Pferdeskelett. *Mitteilungen der Deutschen Archaologischen Instituts Abteilung Kairo* **26**, 43–47.

BOESSNECK, J. & A. VON DEN DRIESCH. 1976. Pferde in 4./3. Jahrtausend v. Chr. in Ostanatolien. *Säugetierkundliche Mitteilungen* **31**, 89–104.

BÖKÖNYI, S. 1968. Data on Iron age horses of central and eastern Europe. *American School of Prehistoric Research, Peabody Museum, Harvard University.* Bulletin **25**, 3–71.

BÖKÖNYI, S. 1974*a*. *The Przevalsky horse* . Plymouth: Souvenir Press.

BÖKÖNYI, S. 1974*b*. *History of domestic mammals in central and eastern Europe.* Budapest: Akadémiai Kiadô.

BÖKÖNYI, S. 1986. The equids of Umm-Dabaghiyah, Iraq. In *Equids in the ancient world* 207–229. R.H. Meadow & H.-P. Uerpmann (Eds). Wiesbaden: Dr Ludwig Reichert.

BÖKÖNYI, S. 1991. Late Chalcolithic horses in Anatolia. I *Equids in the ancient world* II. R.H. Meadow & H.-P. Uerpmann (Eds). Wiesbaden: Dr Ludwig Reichert.

BRUNT, P.A. (transl.) 1976. *Arrian: history of Alexander an Indica.* 2 vols. The Loeb Classical Library. Cambridge, Mass. & London: Harvard University Press.

BUCKLEY, T.A. (transl.) 1853. *The Iliad of Homer.* London: Henry G. Bohn.

BURLEIGH, R. 1986. Radiocarbon dates for human and animal bones from Mendip caves. *Proceedings University Bristol Spelaeological Society* 17, 267–274.

CATLIN, G. 1842. *Letters and notes on the manners, custom and condition of the North American Indians.* 2 vols. London: Tilt and Bogue.

CHOW, B-S. 1989. The domestic horse of the pre-Ch'in period in China. In *The walking larder: patterns of domestication, pastoralism, and predation.* 105–107. J. Clutton-Brock (Ed). London: Unwin Hyman.

CHURCHER, C.S. & M.L. RICHARDSON 1978. Equidae. In *Evolution of African mammals* 379–422. V.J. Maglio & H.B.S. Cooke (Eds). Cambridge, Mass. & London: Harvard University Press.

CLASON, A.T. 1986. Het voorkomen van het wilde paard *Equus ferus* Boddaert, 1785 in Nederland vanaf het Laat-Glacial. *Lutra* **29**, 303–306.

CLUTTON-BROCK, J. 1974. The Buhen horse. *Journal of Archaeological Science* 1, 89–100.

CLUTTON-BROCK, J. 1986. Osteology of the equids from Sumer. In *Equids in the ancient world* 207–229. R.H. Meadow & H.-P. Uerpmann (Eds). Wiesbaden: Dr Ludwig Reichert.

CLUTTON-BROCK, J. 1987. *A natural history of domesticated mammals.* London: British Museum (Natural History) & Cambridge: Cambridge University Press.

CLUTTON-BROCK, J. 1990. Animal remains from the Neolithic lake village site of Yvonand IV, canton de Vaud, Switzerland. *Archives des Sciences Genève* 1, 1–97.

CLUTTON-BROCK, J. & R. BURLEIGH 1979. Notes on the osteology of the Arab horse with reference to a skeleton collected in Egypt by Sir Flinders Petrie. *Bulletin British Museum natural History (Zoology)* 35 2, *127–200.*

CLUTTON-BROCK, J. & R. BURLEIGH 1991. The skull of a Neolithic horse from Grime's Graves, Norfolk, England In *Equids in the Ancient World* vol. 2. R. Meadow & H.-P. Uerpmann (Eds). Wiesbaden: Dr Ludwig Reichert.

CONTAMINE, P. 1986. *War in the middle ages*. Oxford: Basil Blackwell.

COTTERELL, B. & J. KAMMINGA 1990. *Mechanics of pre-industrial technology*. Cambridge: Cambridge University Press.

CROSBY, A.W. 1972. *The Columbian exchange: biological and cultural consequences of 1492*. Connecticut: Greenwood Press.

CUMMINS, J. 1988. *The hound and the hawk*. London: Weidenfeld & Nicolson.

DALQUEST, W.W. 1978. Phylogeny of American horses of Blancan and Pleistocene age. *Annales Zoologici fennici* **15**, 3, 191–199.

DARWIN, C. 1858. *The origin of species by means of natural selection or the preservation of favoured races in the struggle for life*. London: John Murray.

DARWIN, C. 1868. *The variation of animals and plants under domestication*. 2 vols. London: John Murray.

DAVIDSEN, K. 1978. *The final TRB Culture in Denmark a settlement study*. Arkaeologiske Studier Series Monographs **5**, 142–148. Copenhagen: Akademisk Forlag.

DAVIDSON, H.R.E. 1964. *Gods and myths of northern Europe*. London: Penguin Books.

DAVIDSON, H.R.E. 1988. *Myths and symbols in pagan Europe: early Scandinavian and Celtic religions*. Manchester: Manchester University Press.

DENT, A. 1972. *Donkey the story of the ass from east to west*. London: George G. Harrap.

DENT, A. 1974. *The horse through fifty centuries of civilization*. London: Phaidon Press.

DOLLMAN, J.G. 1935. A new race of wild ass from the Sudan, *Asinus asinus dianae*, sub sp. nov. *Proceedings Linnean Society London*. 1934–5, 132–134.

DOUGLAS, D.C. & G.W. GREENAWAY (Eds.) 1953. *English historical documents 1042–1189* Vol 2, pp. 232–78. London: Eyre & Spottiswoode.

DUCOS, P. 1971. Le cheval de Soleb. In *Soleb* **II** *Les Necropoles*, 260–265. M.S. Giogini (Ed.). Firenze: Sansone.

DUNBAR, R. 1985. The race to breed faster horses. *New Scientist*, **1459**, 44–47.

EPSTEIN, H. 1969. *Domestic animals of China*. Farnham Royal, England: Commonwealth Agricultural Bureaux.

EWART, J.C. 1904. The wild horse (*Equus prjevalskii*, Poliakoff). *Proceedings Royal Society of Edinburgh* **24**, 1902–3, 460–467.

EWART, J.C. 1907. On skulls of horses from the Roman fort at Newstead, near Melrose, with observations on the origin of domestic horses. *Transactions Royal Society of Edinburgh* **45**, 555–588.

EWERS, J.C. 1955. *The horse in Blackfoot Indian culture with comparative material from other western tribes*. Smithsonian Institution Bureau of American Ethnology Bulletin **159**.

FITZINGER, L.J. 1857. *Bilder-Atlas zur wissenschaftlich-populären Naturgeschichte der Säugethiere in ihren Sämmtlichen Hauptformen* vol 3, p. 666. Wien.

FLOWER, W.H. 1891. *The Horse a study in natural history*. London: Kegan Paul, Trench, Trübner & Co.

FORBES, H.O. (ed.). 1903. *The natural history of Sokotra and Abd-el-Kuri*. London: R.H. Porter.

FORSTER, E.S. & E.H. HEFFNER (transl.) 1968. *Lucius Moderatus Columella on agriculture*. Vol II. The Loeb Classical Library No. 407. Cambridge, Mass. & London: Harvard University Press.

GAFFNEY, B. & E.P. CUNNINGHAM 1988. Estimation of genetic trend in racing performance of thoroughbred horses. *Nature* **332**, 6166, 722–724.

GEOFFROY, I. ST.-H. 1855. Sur deux chevaux sauvage d'une espèce nouvelle (*Equus hemippus*) donnés par S.M. l'Impératrice à la Menagerie du Museum d'Histoire Naturelle. *Compte Rendus de l'Academie des Sciences* **41**, 1214–1219.

GETTY, R. 1975. *Sisson and Grossman's the anatomy of the domestic animals*. 5th ed. Philadelphia, London etc.: W.B. Saunders.

GILBEY, W. 1903. *Thoroughbred and other ponies with remarks on the height of racehorses since 1700*. London: Vinton & Co.

GLASS, M. 1989. The horse in Neolithic central Europe. Paper presented at the 54th annual meeting of the Society of American Archaeology, April 1989, Atlanta Georgia.

GMELIN, S.G. 1770. *Reise durch Russland*. St Petersburg: Academie der Wikenschasten.

GMELIN, S.G. 1788. *Linnaeus's Systema Naturae*, ed. 13 I pt.I, 213.

GOULD, S.J. 1990. *Hen's teeth and horse's toes*. London: Penguin Books.

GRAY, A.P. 1971. *Mammalian hybrids*. Slough: Commonwealth Agricultural Bureaux.

GRAY, J.E. 1824. A revision of the family Equidae. *Zoological Journal* I *241–248, pl. 9*.

GRAY, J.E. 1850. *Gleanings from the menagerie and aviary at Knowsley Hall*. vol 2. *Hoofed quadrupeds*. Illustrations by Waterhouse Hawkins. Knowsley.

GREENHALGH, P.A.L. 1973. *Early Greek warfare horsemen and chariots in the Homeric and Archaic Ages*. Cambridge: Cambridge University Press.

GRIGSON, C. 1992. The earliest horses in the Middle East? – new finds from the fourth millennium of the Negev. *Journal of Archaeological Science* (in press).

GROVES, C.P. 1963. Results of a multivariate analysis on the skulls of Asiatic wild asses; with a note on the status of *Microhippus hemionus blanfordi* Pocock. *Annals and Magazine of Natural History*. Ser. 13, **6**, 329–336.

GROVES, C.P. 1974. *Horses, asses and zebras in the wild*. London: David & Charles.

GUIDON N. & G. DELIBRIAS 1986. Carbon-14 dates point to man in the Americas 32,000 years ago. *Nature* **321**, 769–771.

HAMILTON SMITH, C. 1845. Horses. In *The naturalist's library. Mammalia*. Vol XX. W. Jardine (ed.). Edinburgh: W.H. Lizars.

HAMMOND, M. (transl.) 1987. *Homer: the Iliad*. London: Penguin Classics.

HANDFORD, S.A. (transl.) 1951. *Caesar the conquest of Gaul* London: Penguin Books.

HARALAMBOS, K.M. 1990. *The Byerley Turk: three centuries of the tail male racing lines*. London: Kenilworth Press, Threshold Books.

HARRIS, W.C. 1839. *The wild sports of southern Africa; being the narrative of an expedition from the Cape of Good Hope, through the territories of the chief Moselekatse to the Tropic of Capricorn*. London: John Murray.

HAYNES, D.E.L. 1975. *An historical guide to the sculptures of the Parthenon*. London: Trustees of the British Museum.

HEDGES, R.E.M., R.A. HOUSLEY, I.A. LAW & C.R. BRONK 1989. Radiocarbon dates from the Oxford AMS system: *Archaeometry* Datelist 9. *Archaeometry* **31**, 2, 207–234.

HEUGLIN, T. VON. 1861. Diagnosen neuer Säugethiere aus Afrika am Rothen Meere. *Novum Actorum. Academiae Caesareae Leopoldino-Carolinae Germanicae Naturae Curiosorum*. **38**. Jenae: F. Frommann.

HILL, W.G. 1988. Why aren't horses faster? *Nature* **332**, 6166, 678.

HILZHEIMER, M. 1941. *Animal remains from Tell Asmar. Studies in ancient Oriental civilization*. No. 20. Chicago: The University of Chicago Press.

HOOPER, W.D. & H.B. ASH (transl.) 1967. *Marcus Porcius Cato on agriculture. Marcus Terentius Varro on agriculture*. The Loeb Classical Library No. 283. Cambridge, Mass. & London: Harvard University Press.

HOUPT, K.A. & A.F. FRASER (Eds.) 1988. Behaviour of Przewalski horses. *Applied Animal Behaviour Science* **21**, 1–2, 1–190.

HSU, T.C. & K. BENIRSCHKE 1967–71. *An atlas of mammalian chromosomes*. Berlin, Heidelberg, New York: Springer-Verlag.

HUNTER, W.P. (transl.) 1837. *Selections from the natural history of the quadrupeds of Paraguay and the River La Plata; comprising the most remarkable species of South America. Translated from the Spanish of Don Felix de Azara. With notes*. London: A.J. Valpy.

HUXLEY, L. (Ed.) 1913. *Scott's last expedition*. 2 vols. London: Smith, Elder & Co.

HYLAND, A. 1990. *Equus: the horse in the Roman world*. London: Batsford.

INGOLD, T. 1980. *Hunters, pastoralists and ranchers*. Cambridge: Cambridge University Press.

JACKSON, K. 1969. *The Gododdin: the oldest Scottish poem*. Edinburgh: The University Press.

JONES, H.L. (transl.) 1983. *The geography of Strabo*. The Loeb Classical Library. Cambridge, Mass & London: Harvard University Press.

KHAZANOV, A.M. 1983. *Nomads and the outside world*. Cambridge: Cambridge University Press.

KIPLING, R. 1902. *Just so stories*. London: Macmillan.

KLINGEL, H. 1974. A comparison of the social behaviour in the Equidae. In *The behaviour of ungulates and its relation to management*. Vol 1, 124–133. V. Geist & F. Walther (Eds). Gland: IUCN Publications N.S. **24**.

LANE FOX, R. 1973. *Alexander the Great*. London: Futura Publications.

LANGDON, J. 1986. *Horses, oxen and technological innovation: the use of draught animals in English farming from 1066–1500*. Cambridge: Cambridge University Press.

LATHAM, R. (transl.) 1988. *The travels of Marco Polo*. London: Penguin Books.

LAWRENCE, E.A. 1984. *Rodeo: an anthropologist looks at the wild and the tame*. Chicago: Chicago University Press.

LAWRENCE, E.A. 1985. *Hoofbeats and society: studies of human-horse interactions*. Bloomington: Indiana University Press.

LEVER, C. 1985. *Naturalized mammals of the world*. London: Longman.

LEVINE, M.A. 1990. Dereivka and the problem of horse domestication. *Antiquity* **64**, 727–40.

LEY, C.D. (transl.). 1965. Overland return from India, 1663. In *Portugese voyages 1498–1663*, 333–358. C.D. Ley (Ed.). Everyman's Library No. 986. London: Dent.

LINNAEUS, C. 1758. *Systema naturae* 10th ed. Vol 1. Facsimile. London: Trustees British Museum (Natural History), 1956.

LITTAUER, M.A. 1969*a*. Bits and pieces. *Antiquity* 43, 289–300.

LITTAUER, M.A. 1969*b*. Slit nostrils of equids. *Zeitschrift für Säugetierkunde* **34**, 3, 183–186.

LITTAUER, M.A. 1971. V.O. Vitt and the horses of Pazyryk. *Antiquity* **45**, 293–294.

LITTAUER, M.A. 1981. Early stirrups. *Antiquity* **55**, 99–105.

LITTAUER, M.A. & J. CROUWEL. 1977. The origin and diffusion of the cross-bar wheel? *Antiquity* **51**, 95–105.

LITTAUER, M.A. & J. CROUWEL. 1979. *Wheeled vehicles and ridden animals in the ancient Near East*. Leiden/Köln: E.J. Brill.

LOCHE, R. & C. SANGER 1988. *Jacques-Laurent Agasse 1767–1849*. London: Tate Gallery London.

LYDEKKER, R. 1904. Notes on the specimens of wild asses in English collections. *Novitates Zoologicae* **11**, 583–596.4 pl.

LYDEKKER, R. 1912. *The horse and its relatives*. London: George Allen.

MAEKAWA, K. 1979*a*. The ass and the onager in Sumer in the late third millennium B.C. *Acta sumerologica*, Hiroshima, **1**, 35–62.

MAEKAWA, K. 1979*b*. Animal and human castration in Sumer part I: cattle (gu_4) and equids (ANŠE.DUN.GI, ANŠE.BARxANo in Pre-Sargonic Lagash. *Zinburn: Memoirs of the Research Institute for Humanistic Studies*, Kyoto University **15**, 95–137.

MAEKAWA, K. & F. YILDIZ 1982. Animal and human castration in Sumer part 3: more texts of Ur III Lagash on the term amar-KUD. *Zinbun: Memoirs of the Research Institute for Humanistic Studies*, Kyoto University **18**.

MARCHANT, E.C. (transl.) 1968. *Xenophon: scripta minora.* The Loeb Classical Library. Cambridge, Mass. & London: Harvard University Press.

MARTIN, P.S. 1967. Prehistoric overkill. 75–120. In P.S. Martin & H.E. Wright, Jr. (Eds.). *Pleistocene extinctions: the search for a cause.* New Haven & London: Yale University Press.

MARTIN, P.S. & J.E. GUILDAY 1967. A bestiary for Pleistocene biologists. In P.S. Martin & H.E. Wright, Jr. (Eds.) *Pleistocene extinctions: the search for a cause.* 1–62. New Haven & London: Yale University Press.

MARTIN, P.S. & H.E. WRIGHT, JR. (Eds.) 1967. *Pleistocene extinctions: the search for a cause.* New Haven & London: Yale University Press.

MARTIN, P.S. & R.G. KLEIN (Eds.) 1984. *Quaternary extinctions: a prehistoric revolution.* Arizona: The University of Arizona Press.

MATSCHIE, P. 1911. Ueber einige von Herrn Dr. Holderer in der südlichen Gobi und in Tibet gesammelte Säugetiere. In K. Futterer *Durch Asien Erfahrungen, Forschungen und Sammlungen.* **III**, 5, 24.

MCDONALD, J.N. 1984. The reordered North American Selection regime and late Quaternary megafaunal extinctions. In *Quaternary extinctions: a prehistoric revolution.* 404–439. P.S. Martin & R.G. Klein (Eds.) Arizona: The University of Arizona Press.

MEAD, J.I. & D.J. MELTZER. 1984. North American late Quaternary extinctions and the radiocarbon record. In *Quaternary extinctions: a prehistoric revolution.* 440–450. P.S. Martin & R.G. Klein (Eds). Arizona: University of Arizona Press.

MOHR, E. 1971. *The Asiatic wild horse.* London: J. A. Allen.

MONOD, T. 1933. Anes sauvages. *La Terre et la Vie.* No. 8, 451–462.

MOORCROFT, W. & G. TREBECK 1841. *Travels in the Himalayan provinces of Hindustan and the Punjab.* London: John Murray.

MOOREY, P.R.S. 1970. Pictorial evidence for the history of horse-riding in Iraq before the Kassites. *Iraq* **32**, 36–50.

MORTON, EARL OF 1821. A communication of a singular fact in natural history. *Philosophical Transactions of the Royal Society* Part I, 20–22.

NAOLA, N. 1970. *On the horse in Japan and East Asia* (in Japanese). Tokyo.

NOACK, T. 1884. Zur fauna des Somalilandes. *Der Zoologische Garten* **15**, 374–5.

NOËTTES, L. DES 1931. L'attelage, le cheval de selle à travers les ages. Paris: Picard.

OLSEN, S.L. 1989. Solutré: a theoretical approach to the reconstruction of Upper Palaeolithic hunting strategies. *Journal of Human Evolution* **18**, 295–327.

OMAN, C. 1924. *A history of the art of war in the middle ages.* 2 vols. London: Methuen.

OPIE, I. & P. OPIE 1951. *The Oxford dictionary of nursery rhymes.* Oxford: Oxford University Press.

OUSTALET, E. 1882. Une nouvelle espèce de le zèbre de grévy (*Equus Grevyi*). *La Nature* Paris **10**, *2nd semestre*, 12–14.

OWEN, A. 1841. *Ancient laws and institutes of Wales.* London: Commissioners of the Public Records.

PALLAS, P.S. 1775. Equus hemionus, mongolis dshikketaei dictus. *Novi Commentarii Academiae Scientarum Imperialis Petropolitanae* **19**, 394–417, Pl. 7.

PETERS, J. 1985–6. Bijdrage tot de archeozoölogie van Soedan en Egypte. Doctoral thesis, Rijksuniversiteit Gent, Falkultiet der Wetenschappen.

PETRIE, W. M. F. 1914. *Tarkhan II.* London: British School of Archaeology in Egypt/Bernard Quarich.

PHARR, C. (transl.) 1952. *The Theodosian code and novels and the Sirmondian constitutions.* Princeton: Princeton University Press.

PIETTE, E. 1906. Le chevêtre et la semi-domestication des animaux aux temps pléistocène. *L'Anthropologie* **17**, 27–53.

PIGGOTT, S. 1968. The earliest wheeled vehicles and the Caucasian evidence. *Proceedings of the Prehistoric Society* N.S. **34**, 266–318.

PIGGOTT, S. 1983. *The earliest wheeled transport from the Atlantic coast to the Caspian sea.* London: Thames & Hudson.

POCOCK, R.I. 1909. On the agriotype of domestic asses. *Annals and Magazine of Natural History.* Ser. 8, **4**, 523–528.

POLIAKOF, M. 1881. Supposed new species of horse from central Asia. *Annals and Magazine of Natural History.* Ser. 5, **8**, 16–26.

POSTGATE, J.N. 1986. The equids of Sumer again. In *Equids in the Ancient World* 194–206. R.H. Meadow & H. P. Uerpmann (Eds). Wiesbaden: Dr Reichert.

POWELL, T.G.E. 1971. The introduction of horse-riding to temperate Europe: a contributory note. *Proceedings of the Prehistoric Society* 37, 2, 1–14.

POWER, M.M. 1990. Chromosomes of the horse. In *Domestic animal cytogenetics* 131–60. R.A. McFeely (Ed.). London: Academic Press.

PROBERT, W. 1823. *The ancient laws of Cambria.* London.

PROTHERO, D.R. & R.M. SCHOCH (Eds.) 1989. *The evolution of the Perissodactyls.* Oxford: Clarendon Press, Oxford University Press.

RAWLINSON, G. (transl.) 1964. *The histories of Herodotus.* 2 vols. Everyman's Library No. 405. London: Dent.

REGALIA, E. 1904. In P.E.Stasi & E. Regalia *Grotta Romanelli (Castro, terra d'Otranto) stazione con Faune interglaciali calda e di steppa.* Firenze: Tipografia 8 di Salvadore Landi.

RENFREW, C. 1987. *Archaeology and language the puzzle of Indo-European origins.* London: Jonathon Cape.

RIDGEWAY, W. 1905. *The origin and influence of the Thoroughbred horse.* Cambridge: Cambridge University Press.

ROGERS, P. 1986. *Daniel Defoe: a tour through the whole island of Great Britain.* London: Penguin Classics.

RUDENKO, S.I. 1970. *Frozen tombs of Siberia the Pazyryk burials of Iron-Age horsemen.* London: J.M. Dent.

RUSSELL, N. 1986. *Like engend'ring like: heredity and animal breeding in early modern England*. Cambridge: Cambridge University Press.

RYDER, O.A. 1988. Przewalski's horse – putting the wild horse back in the wild. *Oryx* **22**, 3, 154–157.

SATTAUR, O. 1991. Rare horses ready for return to Mongolian home. *New Scientist*, 12 January, p. 26.

SAVORY, T.H. 1979. *The mule*. Bushey, England: Meadowfield Press.

SCLATER, P.L. 1884. On some mammals from Somali-land. *Proceedings Zoological Society of London*, 538–542.

SIMPSON, G. G. 1961. *Horses*, New York: Anchor Books Doubleday and The American Museum of Natural History.

SITWELL, N.H.H. 1981. *Roman roads of Europe*. London: Cassell.

SMALL, R.C. 1987. *Crusading warfare 1097–1193*. Cambridge: Cambridge University Press.

SMIELOWSKI, J.M. & P.P. RAVAL 1988. The Indian wild ass – wild and captive populations. *Oryx* *22*, 2, 85–88.

SONDAAR, P.Y. & V. EISENMANN 1989. *L'évolution de la famille du cheval*. Utrecht: Universiteit Utrecht.

SPEED, J.G. & M.G. SPEED 1977. *The Exmoor pony its origins and characteristics*. Chippenham, Wiltshire: Countryside Livestock.

SPONENBERG, D.P. 1984. Preservation of the Spanish horse in North America. *Animal Genetic Resources Information* 333–36. Rome: FAO & UNIP.

SPRUYTTE, J. 1983*a*. *Early harness systems*. London: J. A. Allen & Co.

SPRUYTTE, J. 1983*b*. La conduite du cheval chez l'archer assyrien. *Plaisirs Equestres* **129**. 66–71 [ISSN 0032–051X]

STECHER, R.M. 1962. Anatomical variations of the spine in the horse. *Journal of Mammalogy* **43**, 205–219.

SWAYNE, H.G.C. 1895. *Seventeen trips through Somaliland a record of exploration & big game shooting 1885 to 1893*. London: Rowland Ward.

TANN, J. 1983. Horse power. In *Horses in European economic history a preliminary canter*. 21–30. F. M. L. Thomson (Ed.). Reading, England: British Agricultural History Society.

TEGETMEIER, W.B. & C.L. SUTHERLAND 1895. *Horses, asses, zebras, mules, and mule breeding*. London: Horace Cox.

TELEGIN, D.Y. 1986. *Dereivka a settlement and cemetry of Copper Age horse keepers on the Middle Dneiper*. English transl. by V.K. Pyatkovskiy. Oxford: BAR International Series 287.

THIEBAUX, M. 1967. The mediaeval chase. *Speculum*. **40**, 260–274.

THOMAS, P. 1884. Recherches stratigraphiques et paléontologiques sur quelques formations d'eau douce de l'Algerie. *Mémoires Societé Géologique de France*, Ser 3, 3, 1–50, 4pl.

TREVELYAN, G.M. 1948. *English social history*. London: The Reprint Society.

TRISTRAM, H.B. 1889. *The natural history of the Bible*. London: Society for Promoting Christian Knowledge.

TROW-SMITH, R. 1957. *A history of British livestock husbandry to 1700*. London: Routledge & Kegan Paul.

TROW-SMITH, R. 1959. *A history of British livestock husbandry 1700–1900*. London: Routledge & Kegan Paul.

UERPMANN, H.-P. 1987. *The ancient distribution of ungulate mammals in the Middle East*. Beihefte zum Tübinger Atlas des Vorderen Orients, No. 27. Wiesbaden: Dr Ludwig Reichert.

UERPMANN, H.-P. 1991. *Equus africanus* in Arabia. In *Equids in the ancient world* II. R. H. Meadow & H.-P. Uerpmann (Eds). Wiesbaden: Dr Ludwig Reichert.

ULBRICH, R. 1986. *The great stallion book*. Hobart, Australia: Libra Books.

VAINSHTEIN, S. 1980. *Nomads of South Siberia*. Cambridge: Cambridge University Press.

VESEY-FITZGERALD, B. 1946. *The book of the horse*. London: Nicholson & Watson.

VITT, V.O. 1952. Losadi Pazyrykskich kurganov. *Sovjetskaia Archeologia* **16**, 163–205.

VÖRÖS, I. 1981. Wild equids from the early Holocene in the Carpathian Basin. *Folia Archaeologica* **32**, 37–68.

WARNER, R. (transl.) 1975. *Xenophon: the Persian expedition* Harmondsworth, Middlesex: Penguin Books.

WERTH, E. 1930. Zur Abstammung des Hausesels. *Sitzungsberichte der Gesellschaft Naturforschender Freunde zu Berlin*. Nr. 8–10, 342–355.

WHITE, L. JR. 1962. *Medieval technology and social change*. Oxford: Oxford University Press (Oxford paperbacks).

WIJNGAARDEN-BAKKER, L.H. VAN. 1974. The animal remains from the Beaker settlement at Newgrange, Co. Meath – first report. Proceedings of the Royal Irish Academy **74** *(Section C)* 313–383.

WIRGIN, J. 1985. *The emperor's warriors: catalogue of the exhibition of the terracotta figures of warriors and horses of the Qin Dynasty of China*. Edinburgh: City of Edinburgh Museums and Art Galleries.

YOUATT, W. 1846. *The horse: with a treatise of draught*. London: Charles Knight.

ZARINS, J. 1976. The domestication of Equidae in third millennium B.C. Mesopotamia. Ph.D dissertation no. T.26263, Joseph Regenstein Library, University of Chica

ZARINS, J. 1978. The domestication of Equidae in third millennium B.C. Mesopotamia. *Journal of Cuneiform Studies* **30**, 3–17.

ZARINS, J. 1986. Equids associated with human burials in third millennium BC. Mesopotamia: two complementary facets. *Equids in the ancient world* 164–193. R.H. Meadow & H.-P. Uerpmann (Eds). Wiesbaden: Dr Ludwig Reichert.

ZEDER, M.A. 1986. The equid remains from Tal-e Malyan, southern Iraq. In *Equids in the ancient world*. 366–412. R.H. Meadow & H.-P. Uerpmann (Eds). Weisbaden: Dr Ludwig Reichert.

ZEUNER, F.E. 1963. *A history of domesticated animals*. London: Hutchinson.

ZUBROW, E. 1990. The depopulation of native America. *Antiquity* **64**, 245, 754–65.

Index

Bold figures indicate reference to an illustration or a caption